THE PRACTICE OF MATHEMATICS

Piagetian theory dominates the psychological description and explanation of how children learn to use numbers and do arithmetic. Yvette Solomon presents a comprehensive overview and critique of the field of number development and questions the influential philosophical and psychological underpinnings of Piaget's analysis of the child's conception of number. Her re-interpretation of data produced not only by Piagetians but also by Piaget's critics suggests an alternative approach which has both psychological and educational implications: that knowing how to use numbers entails entering into particular social practices and thus learning how to act appropriately in various contexts.

Yvette Solomon is a Lecturer in Psychology and has taught at Lancashire Polytechnic, W.R. Tuson College of Further Education, and the University of North Wales.

International Library of Psychology

The Practice of Mathematics

Yvette Solomon

ROUTLEDGE
London and New York

First published 1989
by Routledge
11 New Fetter Lane, London EC4P 4EE
29 West 35th Street, New York, NY 10001

© 1989 Yvette Solomon

Filmset by Mayhew Typesetting,
26 St Thomas St., Bristol, England
Printed and bound in Great Britain by
Biddles Ltd, Guildford and King's Lynn

British Library Cataloguing in Publication Data

also available

ISBN 0-415-03038-2

Contents

Contents

Acknowledgements

I would like to thank John O'Neill, Carolyn Stone and Charles Desforges for their invaluable criticism and advice through several drafts of this book. For their help in producing the final manuscript my thanks are due to Sue Merrick, Jenny Girling and Jean Yates for typing various versions, and to Dick Danks for proof-reading the typescript. Finally, I would like to thank my family and friends for their support and encouragement.

1

The development of the number concept as a field of psychological investigation

The central idea of this book involves the argument that current psychological theories of number development share common assumptions about the kind of knowledge that is involved in knowing about number, and that those assumptions can be questioned. The major concern of such theories may be said to be the generation of explanations of how children come to have knowledge, but there is, generally, little discussion of what it is that children know when they know about numbers. This is, apparently, due to the dominance of Piaget's (1952) approach to the psychology of number development; critical discussion centres on Piaget's claims about when children have an understanding of number, and on his account of how this understanding develops, but it rarely challenges his conceptualization of knowing about number itself.

But what a theory says about the development of knowledge is related to what it says, or assumes, about the nature of that knowledge; thus the flaws in Piaget's account of the growth of number understanding may be a result of his assumptions regarding what is involved in knowing number. In Chapters 2 to 5, therefore, I will consider two broad areas: (i) the influences on Piaget's theory which led him to develop the kind of account of knowing number that he proposes; and (ii) his resultant theory of number development in terms of its adequacy as an account of the growth of understanding. Chapter 2 examines the background to Piaget's account and his proposed genetic epistemology, while Chapters 3 and 4 set his theory of the development of the number concept within the context of his epistemological concerns. Chapter 4 also examines Piaget's account of the growth of understanding in terms of its adequacy as a psychological account, while Chapter 5 discusses the general problems of describing and explaining the growth of knowledge in Piaget's theory. I will suggest that such problems are inherent in any theory which, like Piaget's, assumes that knowledge gain is a question of the solitary individual's construction of concepts defined in what I shall call 'essentialist' terms: that is, in terms of the possession of certain logical notions which, it is claimed, are necessary and sufficient conditions for understanding number.

This is an argument which I pursue in Chapters 6 to 8 where I suggest that recent developments in the field, whether critical of Piaget or not, do not show a significant departure from the Piagetian approach. In Chapter 7, I show that attempts to improve on Piaget's theory from within the Piagetian framework do not avoid the problems of his original account; this is also true of Piaget's critics, as I demonstrate in Chapters 6 and 8. Following on from the suggestions made in Chapter 5, Chapter 6 also develops the claim that knowledge is intrinsically social, and that coming to know about number is more fruitfully characterized as entering into the social practices of number use; in Chapter 8 I illustrate this by means of a reinterpretation of the data generated by some recent research into number development and arithmetic skills. Finally, I will make some suggestions in Chapter 9 for a reorientation of the field which focuses on the social practices of number use and an analysis of how children enter into those practices.

The book thus concerns five main themes: (i) the context of Piaget's theory; (ii) the problems of explaining the development of knowledge about number; (iii) the development of the field since Piaget; (iv) a reconceptualization of number understanding; and (v) an analysis of how children are inducted into the social practices of number. For the remainder of this chapter I will consider each of these in turn.[1]

The context of Piaget's theory

Piaget's original intention in studying the development of thought in the child was to answer problems of epistemology concerning the justification of knowledge claims (see Piaget 1966, 1972a, 1972b, for instance). Specifically, in developing an account of the child's conception of number, Piaget intended to contribute to a debate dating back to Kant and concerning the nature and status of mathematical propositions: due to a series of developments in the foundations of arithmetic which are described in the first section of Chapter 2 the question, 'What makes mathematical statements true?' was at the centre of the debate.

For Piaget, the psychological genesis of ideas provides the solution to this question, by transforming it into one which asks, 'What makes mathematics psychologically possible?' His introduction of psychological development into epistemology to form his own genetic epistemology is the subject of the second section of Chapter 2. Piaget argues for a special relationship between logic and psychology in terms of a parallelism between thought structures and the structures of mathematics or logic, while his main aim in genetic epistemology is to synthesize the opposing ideas of mathematics as creation and mathematics as discovery. Here are seen the beginnings of Piaget's psychological accounts of knowing and coming to know as logically structured knowledge arrived at by a process which synthesizes genesis and structure.

Piaget's epistemological concerns lead him to develop a particular kind of theory of the child's conception of number, as Chapter 3 shows: his criticisms of the attempts of the logicist and intuitionist schools to answer the question, 'What makes mathematical statements true?' provide a basis for his account. The first section of Chapter 3 shows how Piaget (1966, 1968a, 1972a) brings empirical evidence to bear on the logicist and intuitionist examinations of the foundations of arithmetic, and judges both to be inadequate on this basis. The result is Piaget's synthesis of order and class described in the second section of the chapter.

In noting the context of Piaget's theory in terms of his intentions and the influences upon him, an important question can be raised regarding the status of the theory as a psychological account. Piaget's theory is clearly a product of his epistemological aims, and as such it lays great emphasis on logical structure and its development. Not only does this lead to problems in explaining development, as I show in Chapters 4 and 5, but it also suggests that Piaget's assumptions regarding the nature of knowing about number should be questioned. The form taken by his theory is not the only possible one, and it may be that, while Piaget is describing what must be true for a judgement to be objective, he does not thereby provide an account of the attainment of that objectivity.

Problems of explaining the development of knowledge about number

Piaget's epistemological concerns extend beyond the account of the foundations of number to a criticism of empiricism and rationalism (1966, 1972a, 1972b) and a restatement of his intention to find a middle way between 'genesis without structure' and 'structure without genesis', which I describe in the first section of Chapter 4. The next section discusses the resultant synthesis of genesis and structure; Piaget is concerned to avoid both the simple mirroring of empiricism and the innate structures of rationalism by means of the assumption of a biological drive towards higher states of equilibrium in which new structures are created from prior ones without being identical to them. Examination of the details of Piaget's (1966, 1968b, 1972a, 1978a, 1978b, 1980) account shows a fundamental problem in the equilibration model which concerns the recognition and resolution of disequilibrium. It apparently begs the question as to how disequilibrium can be recognized in the first place without presupposing the knowledge which is, ostensibly, the product of the resolution of disturbance, not its cause. This is a problem which functions at all levels of development and affects Piaget's account of the separation of subject and object, the formation of concepts, and the completion of operational thought which is necessary for possession of the number concept. As I will show in Chapter 5, Piaget's (1966) emphasis on the solitary abstraction of structure from action as the basis for logico-mathematical knowledge leads to major problems in his theory.

The problems encountered by Piaget's theory are not unique to his account of development. They apply, as I argue in the first section of Chapter 5, to any theory which relies on a model of transition from weaker to stronger logics in order to describe development. Essentially, the problem is that concepts belonging to a stronger logic, if they really do belong to that logic, cannot possibly be generated from the concepts of a weaker logic since, by definition, the stronger logic can express ideas inexpressible in the terms of the weaker. This criticism directly threatens Piaget's account, which relies on an idea of reconstruction and qualitative change in order to achieve its middle way explanation.

Piaget hopes to avoid the threatened collapse into nativism by means of the device of reflective abstraction of new content from the forms of pre-existing knowledge or actions. But, as I argue in the second section of Chapter 5, his notion of reflective abstraction fails to present a real alternative to ordinary abstraction because it is not possible that someone could abstract the concept of, say, one–one correspondence from the activity of playing with pebbles without the concept of one–one correspondence being presupposed. Furthermore, while the action that the child engages in in playing with pebbles could be described as that of putting things into one–one correspondence, this does not necessarily mean that the child sees it as that action, or could ever see it as that action while he remains outside of a mathematical context.

If the meaning of action is given by the context in which it occurs, then understanding of number use cannot be abstracted from action by the individual alone; in the third section of Chapter 5 I argue that Piaget's equilibration model cannot account for the growth of knowledge that is objective. Thus I present the first part of the argument that knowledge is intrinsically social and that the development of number understanding must therefore be reconceptualized as entering into social practices: understanding numbers is a question of knowing how and when to use them appropriately in different contexts. This argument is pursued in Chapter 6.

Development of the field since Piaget

Piaget's general approach tends to dominate the field of investigation into development of the number concept; even theorists who are critical of Piaget appear to share his assumptions regarding the nature of understanding number, as Chapters 6 and 8 show. Piaget's account of concept possession is an essentialist one: it assumes that all uses of number have a common factor and that the term 'number' has a unitary meaning such that understanding of the number concept can be attributed to someone if they fulfil certain necessary and sufficient conditions, namely, understanding of the logical notions of class, order, and one–one correspondence. In the first section of Chapter 6 I will criticize this account of number understanding and argue instead that knowing about numbers involves a series of applications of number in different situations, and that this knowledge can be shown in a variety of ways, through a variety of behaviours.

Although theories of number development since Piaget have followed new trends, for instance the information-processing approach or the move into arithmetic, an essentialist account of concept possession remains. Information-processing models represent an attempt to express the dynamism of development which is missing in Piaget's theory by means of the statement of knowledge in procedural rather than declarative terms. But as the second section of Chapter 7 shows, Case's (1978a, 1978b, 1982, 1985) attempt to fill the gaps in Piaget's account of growth results in a theory which relies on a basic abstraction model of development in which the abstraction of new knowledge presupposes the very knowledge to be gained. Its effect is more to emphasize the social nature of knowledge than to produce an adequate account of its development. In the first section of Chapter 7 I discuss another recent development of Piagetian theory; this is Doise and Mugny's (1984) and Perret-Clermont's (1980) attempt to introduce social factors into the Piagetian account of development by means of a socio-cognitive model of conflict. They do not, however, produce a theory which answers the criticisms of Piaget's original account: their model of conflict introduces social factors into the growth process, but fails to recognize the essentially social nature of knowing itself. Furthermore, there is reason to doubt whether children really experience the conflict on which Doise, Mugny and Perret-Clermont's model relies; this is also a criticism of Case's information-processing version of Piaget's theory.

Essentialism remains in the accounts of number development given by Piaget's critics too: in the second and third sections of Chapter 6 I discuss two theories, Bryant's (1972, 1974) and Gelman and Gallistel's (1978), in these terms. Even though it opposes Piaget's theory, Bryant's account shows basic similarities: it assumes that having the number concept entails having access to essential logical principles and that the possession of these can be demonstrated by success in any task isomorphic in structure to those principles. Gelman and Gallistel's account also opposes Piaget in that their initial aim is to give a positive account of what children actually know, rather than a negative one couched, as Piaget's is, in terms of logical lacunae. They do not, however, fulfil this purpose, ending with an account which merely sets a lesser criterion for having the number concept while retaining the Piagetian concept of knowing. For Gelman and Gallistel, failure on a test is the result of a failure of cognitive organization, a lack of 'algebraic reasoning', or the overwhelming nature of certain experimental materials; their account assumes that children understand what is required of them in the experimental situation. Relatedly, Bryant's theory also assumes that the part played by context in knowledge is unimportant, so that a task isomorphic in structure to a particular logical principle can be solved solely by reference to that structure with no other information or understanding necessary; the possibility is not considered that understanding of the situation itself in terms of participating in its social practices is in fact intrinsic to success in the task in question.

This is also true of the work reported by Donaldson (1978) and her co-workers, and discussed in the first section of Chapter 8. These researchers maintain that the design of Piagetian tasks is such that they produce 'false negative results': linguistic, non-linguistic, and social factors contribute to masking the child's true competence. Thus Donaldson's main conclusion is that children are unable to show their competence because they cannot attend to the language of the test question and so 'disembed' it from the non-linguistic aspects of the situation. Defences of Piaget, however, claim that the modified tasks designed by Donaldson and her co-workers simply produce 'false positive results': they allow children to respond correctly without recognition of the logical necessity of their answers. Both of these views, however, assume that correct solution of Piagetian tasks demands a certain underlying logical competence. They fail to capture the relationship between children's behaviour in such situations and their interpretation of the experimental context and the way in which language is used in it to indicate the experimenter's intention. In particular, while Donaldson does pay some attention to the child's interpretation of the experimental situation, she assumes, contrary to the position I develop in the first section of Chapter 6, that language understanding can be separated from the context in which it is used and that correct solution of Piagetian tasks is a question of attending to 'pure linguistic form'. An alternative view is that, if the meaning of words is given by their use, then their interpretation changes according to the situation in which they appear; knowing the appropriate way to respond in a Piagetian task requires that someone should be familiar with the social practices associated with the test situation.

Reconceptualizing number understanding

An analysis of the problems encountered in describing the development of knowledge about numbers, together with a reinterpretation of data generated in number and arithmetic research, suggests the possibility of a reconceptualization of knowing about number as entering into the social practices of number use. In the first section of Chapter 6 I criticize Piaget's essentialist account of concept possession for its assumption that the term 'number' has a unitary meaning and that it is possible to specify in abstract the necessary and sufficient conditions for having the concept of number. I argue that it is impossible to specify *a priori* the conditions for knowing the meaning of numbers since this meaning is embedded in the behaviour that surrounds their use. Thus if a person is to be said to understand numbers, they must be able to do a variety of things with numbers; they must know how to act according to norms of appropriateness in a variety of situations to do with using numbers.

Thus I will argue that understanding is intrinsically social; knowing about number entails knowing how and when to use and respond to numbers according to the context in which they appear. Researchers concentrating on 'pure tests' for the number concept assume that children understand what is required of them in the experimental situation, or that, if children possess the essential principles of number, then they can easily and correctly apply numbers in situations of which they have little or no experience. An adequate reconceptualization of number understanding entails that, in assessing children's performance in tests, it is necessary to take into account how far a child can be said to have entered into the various social practices associated with using numbers.

The function of context becomes very clear in research into children's understanding of arithmetic, an idea which is initially raised in the second section of Chapter 7. Thus in the second section of Chapter 8 I argue that understanding of arithmetic is something which does not come about of its own accord. Number use is not necessarily abstractable, self-evident or inevitable; there is no necessary reason why a child should abstract the understanding of arithmetic from his everyday play with objects and use of numbers, for they may not be related in any way. According to this account, understanding arithmetic involves understanding the context of arithmetic, or, in other words, entering into its social practices.

This view is supported by a reinterpretation of data such as that supplied by Steffe *et al.* (1982, 1983), which I discuss in the third section of Chapter 8. Steffe *et al.*, like Piaget, consider mathematical knowledge to be an individual construction, and they are concerned to match children's 'conceptualisation of numerical units' with their problem-solving abilities. But a close examination of Steffe *et al.*'s (1982) protocols suggests that children's performance on arithmetic problems is dependent on their understanding of the context of doing arithmetic, not on their level of conceptual knowledge. Specifically, much appears to turn on the question of what arithmetic is about – hypothetical problems as an end in themselves – and the methods it uses – representations, rules for their manipulation, and their relationship to the problem.

Entering into the social practices of number use

I will argue, then, that knowing about numbers should be reconceptualized as having knowledge that is intrinsically social: on the basis of a number of arguments, and particularly that put forward in the first section of Chapter 6, I contend that such a reconceptualization generates a more coherent explanation of how a child comes to know. Thus knowing about number can be described as entering into the social practices of number use, and coming to know as a process of initiation into a social understanding of when and how to act in particular situations. In Chapter 9 I shall discuss the reorientation of the field that such a reconceptualization suggests, and the methods of data collection and interpretation that it requires, drawing on two related areas of investigation: observations of number use at home and in the classroom, and studies of language and learning.

2

Why does Piaget's theory take its particular form?

Piaget's theory of number development in the child as first stated in psychological terms in 1952 and subsequently elaborated in philosophical terms (in, for instance, Beth and Piaget 1966) did not occur in isolation from other thinking on the subject of mathematics. In particular, Piaget draws on a wider tradition of thought dating back to Kant and concerning the nature and status of mathematical propositions; a little recognized feature of Piaget's work is that it has an epistemological aim. That is to say, one of Piaget's concerns is to comment on the role and development of mathematical thinking within the wider body of human knowledge. Thus Piaget's original intention in studying the development of thought in the child was to answer problems of epistemology concerning the justification of knowledge claims. Specifically, Piaget claims that:

> Man cannot understand the universe except through logic and mathematics, the product of his own mind; but he can only understand how he has constructed mathematics and logic by studying himself psychologically and biologically, or in other words, as a function of the whole universe.
>
> (1972b: 83)

Thus in constructing a psychological theory of number development, Piaget makes reference to an earlier debate concerning the nature and status of mathematical propositions. His psychological theory is intended to be a contribution to this debate:

> In the study of the development of intellectual operations, developmental psychology describes the formation of logico-mathematical structures in terms of a formalisation which is drawn from logic itself. Thus there already exists an exchange of knowledge between logicians and psychologists on the subject of the developmental relationship and the formal genealogy of structures of this kind. One example is the formation of the concept of natural numbers.
>
> (1972b: 101)

9

In this chapter I will examine Piaget's reasoning in the construction of his psychological theory of number development. The first section provides an examination of the historical background to the theory, from Kantian thought to the twentieth-century developments in the foundations of mathematics. The second section considers Piaget's theoretical rationale for bringing together logic and psychology to form his own genetic epistemology.

The background to Piaget's explanation: the status and nature of mathematical propositions

Piaget's reasoning is most clearly seen as the result of his attempt to contribute to the debate concerning the status and nature of mathematical propositions. This debate concerns the apparent necessity of certain statements, for instance '1 + 1 = 2', 'any triangle which is equiangular is also equilateral', or 'if this animal belongs to the class of cats and if this class is included in the class of vertebrates, then this animal belongs to the class of vertebrates'. All these statements appear to be true by necessity. It is the examination of such necessity which forms the basis of the philosophy of mathematics. Thus, as Körner (1960) puts it:

> Why is it that they appear to be necessarily, self-evidently or indubitably true? Are they true in this peculiar way because they are asserted about objects of some special type – namely numbers, shapes, classes; or because they are asserted about objects in general or 'as such'; or are they possibly true in this special way because of their not being asserted of any objects at all? Is their truth due to the particular method by which they are reached or are verifiable – for example, an immediate and incorrigible act of intuition or of understanding?
>
> (1960: 10)

Although philosophers as far back as Plato attempted to answer these questions, a selective historical account relevant to Piaget's theory begins with Kant and reaches into the nineteenth and early twentieth centuries. In generating his own particular answer to the question of 'what makes mathematics true?', Piaget makes reference to many of the earlier doctrines while setting up his own theory in direct opposition to them. In this first section I will briefly outline these earlier developments.

Kant: *mathematics as synthetic* a priori *knowledge*

Prior to Kant, philosophers had treated propositions as falling into one of two categories: truths of reasoning, which are necessarily true by virtue of their logical form, such that a contradictory statement would not be consistent; and truths of fact which are contingent and can have consistent contradictions. Kant (1933, 1977), however, divided propositions into three types, not two: he replaced the dichotomy between analytic and factual propositions with a threefold distinction between analytic propositions, synthetic *a posteriori* propositions, and synthetic *a priori* propositions. Analytic knowledge, which must be *a priori*, is trivially true: negation of such a proposition results in a statement which is contradictory; however, synthetic knowledge presents the problem of how we can know that synthetic propositions are true: the concepts they link are not intrinsically related. For synthetic *a posteriori* statements, the synthesis of separate concepts depends on sense experience. Synthetic *a priori* propositions, on the other hand, take their necessity from the fact that they must be true if any proposition about the physical world is to be true: propositions about external objects concern both the empirical material located in space and time, and the space and time itself; space and time are the conditions necessary for objective experience in that they contain the material for perception, but are not part of it. Thus Kant felt that the clearest instances of synthetic *a priori* knowledge were provided by mathematical propositions, which describe space and time: Euclidean geometry is about space and arithmetic is about time. They are synthetic, because they describe unrelated concepts, but they are also *a priori*, because they do not depend on sense experience.

While commentators (for example, Rotman 1977) have noted that there are many similarities between the general ideas of Kant and Piaget, the material for Piaget's work on the psychology of number comes largely from later developments in the philosophy of mathematics. The rest of this section describes these developments.

The development of non-Euclidean geometry

Until the nineteenth century, mathematicians were generally agreed that geometry, in the form of Euclid's postulates, was knowledge that was necessarily true: any negation of the postulates was inconsistent. Doubts began to emerge, however, with the appearance of non-Euclidean geometries, which were developed as a result of concern about Euclid's fifth postulate which was, supposedly, a self-evident law. By comparison with the other four, the fifth postulate appeared to be too complex and less clearly true, and mathematicians therefore tried to eliminate it, by deducing it from the others and thus showing that it was not an independent postulate, or by simplifying it to something more obviously self-evident. Although these attempts failed, they had the unexpected effect of showing that there were other principles that could be substituted for the fifth postulate and still allow the deduction of theorems: the result was the development of non-Euclidean geometries.

Although it was felt that the postulates and theorems of the non-Euclidean geometries must necessarily be false, and therefore that the systems must be logically inconsistent, this could not be proved; on the other hand, non-Euclidean geometries could not be shown to be consistent either, and so mathematicians were prompted to look for more rigorous procedures than Euclid had used.

These new procedures involved presenting proofs which were valid solely by virtue of their logical form, not by appeal to self-evidence. This meant ignoring the meanings of the primitive terms and so producing an uninterpreted system, in contrast to Euclid's interpreted one. To demonstrate the consistency of such a system, one can: (i) interpret the system in such a way that all the axioms are true (which relies on the ability to judge the truth of the interpreted statements); or (ii) show that one system is consistent if another, less suspect one, is. Using the latter method, mathematicians were able to show that non-Euclidean geometries are consistent if Euclidean geometry is. But the consistency of Euclidean geometry could no longer be assumed on the basis of the synthetic *a priori* truth of the postulates. To establish consistency for Euclid's system, mathematicians took the uninterpreted version and gave the system a numerical interpretation using real numbers. Thus the consistency of Euclidean geometry depended on that of number theory.

The foundations of arithmetic

Before the nineteenth century, arithmetic was never axiomatized in the same way as geometry had been, but, rather, took the form of a collection of rules for calculation. This situation changed when the nineteenth-century mathematicians developed a theory of numbers whereby mathematics could be generated from the natural numbers. This arithmetization was felt to be necessary because of doubts over the legitimacy of the higher types of numbers (that is, real and rational), on which analysis depended. Analysis itself was shown to depend on a good deal of intuition, the reliability of which was being thrown into doubt.

One–one correspondence, set theory and transfinite numbers

The natural numbers were axiomatized by Peano (1889) in five axioms, together with three undefined terms: 'zero', 'immediate successor', and 'natural number'. Peano's system expressed the essential principles of the natural numbers, but in order to generate higher numbers from these it is in fact necessary to add the two terms 'ordered pair' and 'set', which are notions from the theory of sets. A further development of the arithmetization of analysis was Cantor's (1895–7) theory of sets and transfinite numbers. A fundamental notion in the theory of sets is that of one–one correspondence, which denotes the equivalence of two sets. Two sets are said to be in one–one correspondence if there is a relation that holds between them such that each member of the first set is correlated with one and only one member of the second set, and each member of the second set is correlated with one and only one member of the first set.

The notion of one–one correspondence can be used to compare infinite sets as well as finite sets. Thus the set of natural numbers 0, 1, 2, 3 . . . can be put into one–one correspondence with the set of the odd numbers, 1, 3, 5, 7. . . . Cantor's theory went further than this to claim that not all infinite sets were the same size; in particular, the set of real numbers cannot be put into one–one correspondence with the set of natural numbers. In fact, Cantor showed that not even the subset of real numbers between 0 and 1 could be put into one–one correspondence with the natural numbers. A further development of these ideas was to create a theory of transfinite cardinal numbers, numbers which measure infinite sets, so that the set of natural numbers could be given a cardinal number, which was the smallest transfinite number, as could the set of real numbers, whose cardinal number is large. Cantor's work was seen by a number of mathematicians and philosophers as the key to the development of a single axiom system from which could be generated the whole of mathematics, including geometry, which, as we have seen, was given a numerical interpretation in order to secure its consistency.

The intuitionists

An important consequence of Cantor's theory of transfinite numbers was that it countered the views of philosophers such as Aristotle and Kant, who had stated that there cannot be such a thing as an actual infinity. For Kant, synthetic *a priori* knowledge of the laws of number is possible in the sense that the mind is gaining insight into its own inner workings, not into reality itself. It follows from this that there cannot be an actual infinite, only a potential infinite. Kant's philosophy was taken up by a group of mathematicians subsequently called the intuitionists, who saw the intuition of temporal counting as the basis of number. Their argument against Cantor's transfinite numbers depended on their claim that methods in mathematics must involve a finite number of steps. Cantor's proof that the set of real numbers is larger than the set of natural numbers relies on a rule that requires an infinite number of steps. Thus the intuitionists rejected Cantor's proof as 'non-constructive', and with it the whole theory of transfinite numbers, requiring instead that before a mathematical statement about numbers can be said to be true, we must have a constructive proof of it. The intuitionists also rejected the law of the excluded middle, that is, that every statement is either true or false:

> The intuitionist believes that numbers are creatures of the mind, and he believes with Kant that whatever the mind creates it must in principle

13

be able to know through and through. He holds that there can be no unknowable (that is, not constructively provable) truth or falsity about numbers. Therefore he maintains that we have no assurance that assertions that have been neither proved nor disproved are either true or false. If they can be neither proved nor disproved, then they are neither true nor false.

(Barker 1964: 76)

The intuitionist view of mathematics was not widely accepted; the major problem with the intuitionist stance was that it implied the jettisoning of large parts of useful classical mathematics.

The logicist school

Realist mathematicians did not make the criticisms of Cantor's theory that intuitionism makes. For realists, mathematics is a question of discovery, not creation. Frege (1884), for instance, maintained that knowledge of numbers was a question of *a priori* rational insight, and was analytic by virtue of its logical form. Frege's view that the laws of number are reducible to the laws of logic was an important innovation, and represented the basic tenet of the logicist school. Russell (1919) independently had similar ideas to Frege, and with Whitehead (Russell and Whitehead 1910–13) attempted to formulate the logistic thesis explicitly. In so doing, they needed to: (i) state clearly the laws of logic; and (ii) define the basic notions of number theory in such a way that its laws could be deduced from logic. Such a formulation required a more powerful system of logic than had hitherto existed; Frege, Russell and Whitehead developed this and also included as part of logic the set-theoretical notions of 'set' and 'ordered pair'.

The basic notions of number theory – 'zero', 'immediate successor', 'natural number', '+' and '×' – all needed new definition. Barker's version is as follows:

zero is the set of all empty sets

one is the set of all non-empty sets each of which is such that any things belonging to it are identical

two is the set of all sets each having a member distinct from some other member but each being such that any member is identical with one or other of these

a set of sets is the immediate successor of another set of sets if and only if, when one member is removed from any set belonging to the former, then the diminished set belongs to the latter.

(1964: 80)

The natural numbers can be defined as 'anything belonging to every set to

14

which zero belongs and to which belongs the immediate successor of anything that belongs' (1964: 81). Together with the definitions of addition, multiplication, sets, and ordered pairs, this system can generate Peano's axioms and the rest of number theory. While Frege thought that the logistic thesis could only apply to number theory, Russell and Whitehead held that all mathematics could be reduced to logic.

The paradoxes in set theory

The logistic thesis received a severe setback with the discovery by both Cantor (1899) and Russell (1902) of paradoxes in the basic assumptions of set theory as it is stated informally. Cantor's paradox was a product of asking the question: 'Has the entire set of cardinal numbers a cardinal number?', to which the reply, to fit with Cantor's basic assumption that every set has a cardinal number, must be yes. But, on the other hand, the idea of such a number violates the essential principle of the theory of transfinite numbers: that there cannot be a largest number, for the number of all the cardinals must be larger than any cardinal number.

Russell's paradox revealed a more basic problem in set theory. It showed that a basic axiom of set theory involved in Frege's attempt to reduce mathematics to logic – the axiom of abstraction – led to a contradiction. The axiom states that it will be true, however we fill the blank, that:

> There exists a set such that whatever x may be, x is a member of it
> if and only if . . . x . . . (for example, x is a horse).

But if the blank is filled with 'x is a set that is not a member of itself', then we get a sentence that states the existence of a set of all sets that are not members of themselves. On the premise that either this set is a member of itself or it is not, a contradiction arises: if it is a member of itself, then it cannot fulfil the conditions for membership of the set, so therefore it is *not* a member of itself. On the other hand, if it is not a member of itself, then it fulfils the condition for membership of the set, so it *is* a member of itself.

Since set theory was so essential to the arithmetization of analysis and the logistic thesis, it was imperative to restore consistency and thus ensure the reliability of mathematics. Russell argued that the basis of the paradoxes lay in their common violation of what he called 'the vicious circle principle', which, informally stated, treats as meaningless definitions of x which refer to a totality to which x belongs. Cantor's paradox violates this because it defines the number of the cardinal numbers by referring to the set of all cardinals, to which the number belongs. Similarly, Russell's paradox defines the set of all sets not members of themselves by reference to the set of all those sets, to which the set under definition belongs.

In order to avoid the paradoxes, Russell and Whitehead developed the theory of types, with the idea of expressing the vicious circle principle formally. The aim of the theory of types was to differentiate the referents of set theory into a hierarchy of types, so that, for instance, sets, sets of sets, sets of sets of sets, etc. are all distinct from each other. The hierarchy begins at the lowest level with individuals, goes to the second level which contains sets whose members are entities of the lowest type, and so on, until sets of entities of the nth type belong to the type n + 1. Sets having members of types apart from the next lowest type are regarded as meaningless, and it is this constraint that prevents the paradoxes from arising. The condition 'x is a set that is not a member of itself' is rejected as meaningless.

However, the introduction of type theory resulted in the necessity for increased complexity in the theory of sets; for example, it was necessary to introduce various axioms, for instance the axiom of infinity, which threw doubt on the extent to which Russell and Whitehead could be said to have deduced mathematics from logic.

The formalists

The paradoxes raised the issue of consistency again, and with it the question as to whether the consistency of a revised version of set theory could be assured. The two previously used methods of testing consistency – finding an interpretation under which all the axioms come out as indubitably true, or finding a relative consistency proof using some less suspect system – could not yield convincing results since relative proof could not progress very far, and an interpretation involving reference to infinite arrays would be highly suspect with regard to truth.

Hilbert (1904, 1925, 1927), the major figure of the formalist school, proposed a non-relative method which treated all the symbols or terms in a system as completely meaningless, so that the only interest in the system is in the way that strings of marks go together to make sentences. Hilbert's metamathematics describes a system first in terms of formation rules that say which are and which are not permissible combinations of marks and therefore well-formed formulas. Some of the latter will be theorems of the system. The system so described can be studied with complete disregard for the meanings of any of the marks; however, statements about the system are made in a metalanguage, and constitute metamathematical study, which *does* take meaning into account. A system so formalized can be examined for consistency by means of metamathematics, which can be used to find out if contradictions exist. Metamathematical reasoning itself needs to be reasonably above suspicion if it is to be used to establish consistency. For this reason, Hilbert made the restriction that it must use only constructive methods – a metatheorem must entail a method involving a finite number of steps for demonstrating its claim.

As well as consistency, it was also considered important that a system should have completeness, 'the idea that nothing which ought to be a theorem of the system fails to be provable as a theorem' (Barker 1964: 94). Metamathematics can be used to examine completeness as well as consistency, and Hilbert and others in the formalist school hoped to establish a single axiom system for the whole of mathematics which was both consistent and complete. However, Gödel (1931) demonstrated that these two qualities were mutually exclusive: with regard to the natural numbers at the root of the question of foundations, Gödel was able to show, by means of the system of 'Gödel numbering', that a formula existed that is not a theorem if it expresses a truth about the natural numbers, and is a theorem if it expresses a falsehood about the natural numbers; the existence of this formula means that the system is inconsistent if it is complete, and incomplete if it is consistent. It was also a consequence of Gödel's result that a metamathematical description of a system must use a richer logic than that of the system itself, so that Hilbert's hope of using only constructive proofs could not be fulfilled.

Following Gödel's proof in 1931, the state of the debate on the foundations of mathematics into which Piaget first entered a decade later was as follows: mathematicians had attempted to answer the question, 'What makes mathematical statements true?'. For the intuitionists, mathematical propositions were synthetic *a priori*; numbers arise from the pure intuition of time, and constructive proofs ensure the truth of statements. Intuitionism, however, was criticized for losing much of classical mathematics as a result of the demand for constructive methods. For the logicists, *a priori* analytic truth lay in the logic from which the laws of number theory could be deduced. Part of this logic contained set-theoretical notions. However, the paradoxes of set theory created such complications for the logicist programme that its credibility was highly questionable. Finally, formalism made no claim to reinstating the truth of mathematics itself, its statements being treated as meaningless and as having no truth-value. On the other hand, the formalist account held that truth lay in the statements of metamathematics. But the programme collapsed after Gödel's discovery of the mutual exclusivity of consistency and completeness.

Piaget's work in mathematics aims at contributing to the debate concerning the status and nature of mathematical propositions. Critical of earlier attempts to solve the problem, Piaget (1966, 1972a) argues that the psychological genesis of ideas provides the key to a solution. Thus the question, 'What makes mathematical statements true?' becomes 'What makes mathematics psychologically possible?'. His position, Piaget claims, solves the problem of accounting for the appearance of necessity of mathematical statements without encountering any of the problems found in earlier solutions. His account at the same time draws on the positions outlined in this section, and can in particular be seen as a synthesis of the logicist and intuitionist accounts. Thus his theory of the development of the number concept is not, and is not

17

claimed to be, purely psychological. It is intended as a contribution to epistemology, and this intention has important consequences for his account as a contribution to psychology. The next section outlines Piaget's rationale for introducing psychological data into the foundations of mathematics. His criticisms of logicism and intuitionism and his resultant theory of number development follow in Chapter 3.

Mathematics and genetic epistemology: the relation between logic and psychology

Piaget's (1966, 1972a) approach to the question of how to account for the apparent necessity of mathematics is a radical one: he introduces psychological development into epistemology and so proposes to solve many of its problems. In so doing, Piaget also sets himself outside of the debate just described and analyses the existence of the debate itself, from Plato to Gödel, in terms of the psycho-historical development of formalization. Piaget's approach thus translates the question, 'What makes mathematical statements true?', into 'What makes mathematics psychologically possible?'. These two are equivalent for him by virtue of the arguments contained in genetic epistemology, which state a special relationship between logic and psychology.

Genetic epistemology involves a psychological interpretation of mathematics or logic such that the main question to be answered as far as Piaget is concerned is one which asks why certain laws impose themselves on the individual. As I will show in this section, Piaget answers in terms of a parallelism between thought structures and the structures of mathematics or logic. His main aim in genetic epistemology is to synthesize the opposing ideas of mathematics as discovery and mathematics as creation. To this end he claims that mathematical thought is the result of action on objects, which itself is the result of certain intellectual tendencies to act in particular ways. Mathematical knowledge is abstracted from action on objects rather than from the objects themselves, and thus is constructed by the subject while at the same time being rooted in physical experience. The psychological possibility of pure mathematics involves a progress towards the symbolic carrying out of actions without reference to the original objects of the actions.

From a theory of the development of mathematical thought in the individual, Piaget goes on to give an account of the relationship between the development of thought in the individual and the historical development of mathematics. Within this context, he also extends his analysis to include an account of the nature of logic and of the limitations in logic discovered by Gödel, and finally he argues the merits of the formalization of ordinary thought in genetic epistemology. Discussion of Piaget's account shows some self-contradiction in his rationale for genetic epistemology, and raises the question as to whether Piaget's particular formalization of thought is adequate to account for, or describe, the development of knowledge.

The relation between the structures of thought and logic

If epistemology is the study of the grounds on which, and method by which we can say that we know something to be true, then, to use an earlier example, epistemology is concerned with why and how it is that the statement '1 + 1 = 2' appears to be necessarily true. It is thus distinguished from the mathematical study represented by '1 + 1 = ?'. Finding the result of one plus one is a problem for the mathematician while finding the reason for the apparent necessity of the answer being two is the domain of the epistemologist. Psychologism is the use of psychology as a substitute for deduction in order to solve logical or mathematical problems of the order '1 + 1 = ?'.

By asking the question, 'What makes mathematics psychologically possible?', Piaget exposes himself, he thinks, to the charge that his theory is psychologistic (see Piaget 1966, for example). He is anxious not to be so accused and states repeatedly his idea that psychology can only enter into the field of logic when it takes as its problem the determination of the mental mechanisms by which logical problems and their solutions develop in the mathematician's mind. Strictly speaking, though, Piaget need not defend himself against a charge of psychologism, since this would not apply to epistemology anyway, because, as I have stated, psychologism is the application of psychology to the mathematical problem '1 + 1 = ?', rather than to the epistemological problem, 'What makes "1 + 1 = 2" apparently true?'. Psychologism is more accurately the equating of logical laws with the laws of thought, as is the case with the work of the nineteenth-century logician Boole. Thus to apply psychological data to the solution of the question, 'What makes logic true?' is not psychologistic. It is, however, an epistemological stance criticized by many as an example of the 'genetic fallacy' (notably Hamlyn 1967, 1971, 1978). It is considered fallacious because it is not necessarily the case that epistemological questions concerning the nature and status of a proposition (that is, what makes something true) can be answered by reference to psychological facts about the way that someone comes to know. There are two aspects to the matter: (i) genetic priority does not necessarily imply epistemological priority (that is, if one learns P by learning Q, it does not necessarily follow that knowing Q is a necessary condition for knowing P); and (ii) how a person comes to know something is independent of any justification of the truth of what is known. However, this criticism of Piaget is not my main concern here.

For Piaget, genetic epistemology involves:

Looking for the psychological explanation which, as we shall see, will continue to transform itself into a kind of psychological correspondence of the position which the logician is led to adopt, by virtue of the autonomous development of research into the foundations [of mathematics] . . .. By introducing . . . a radical separation between questions of validity or norms and those of fact or genesis, we may give a psychological interpretation of mathematics or logic which does not simply consist in discussing them but attempts to understand in terms of genetic processes how such constructions may be explained, including those which are relevant to the foundations of these subjects.

(1966: 135)

Elsewhere, Piaget states his case more strongly:

To take a trivial example, if the truth of $2 + 2 = 4$ is not a factual datum but a logical demonstration it nonetheless remains true that the epistemological problem is not solved when we show why the demonstration is valid: we still have to know what 2, 4, + and = 'are' or 'designate', and what the subject does to comply with the normative necessity of this demonstration. To say that 2, 4, + and = are rational entities and that the subject does not play a part in their organisation . . . is one epistemological solution amongst others, but others are possible which also respect the same normative autonomy of this demonstration. In order to choose between these solutions, the psychological data always and necessarily enter in. The coordination of factual data and normative validities will therefore consist in putting them into correspondence without reducing one to the other.

(1966: 152–3)

As far as Piaget is concerned, epistemology by definition ought to be concerned with psychological facts. The fact that we accept the rules of logic as valid and apply them in behaviour or thought is a psychological fact, and Piaget's epistemological concern is to examine why certain laws impose themselves on the subject. To return to mathematical propositions and Körner's question, 'Why is it that they appear to be necessarily, self-evidently or indubitably true?', Piaget answers, 'Because there is a parallelism between the structures of thought and the structures of mathematics or logic'. This parallelism is the central core of Piaget's theory; it provides the rationale and method of genetic epistemology and leads to an overall theory of knowledge which includes mathematical epistemology:

The problem which we deal with here is thus central to the object of this work: to explain psychologically why pure mathematics has become possible, is, in short, to declare oneself for or against a connection

20

between logico-mathematical entities and the subject's activities . . .
The examination of the problem must therefore begin with an analysis
of logico-mathematical experience at its most elementary levels, so as
to decide whether this experience is reduced to one relating to objects
in the sense of external physical experience; or is concerned with states
of consciousness (that is, the subject considered as an object) in the
sense of internal psychological experience; or again whether it is a
question of another type of experience concerned with the result of
actions and their coordinations, that is, which consists in noting the
result of these combinations in the way we note the results of a calculation, which allows us sooner or later to replace experience by an operational deduction.

(1966: 231)

The synthesis of mathematics as discovery and mathematics as creation

Piaget's aim, then, is to steer a course between empiricism and rationalism
and between mathematics as discovery and mathematics as invention to end
with a picture of mathematics as at the same time constructed by the mind
and also rooted firmly in the physical universe. This is possible because
logico-mathematical thought is the result, not of direct experience of the
world, but of our actions in the world. Knowledge is abstracted from these
actions and not from the physical properties of the object. So, to take an
example from Piaget, suppose a child plays with pebbles, and counting ten,
discovers that there are still ten even when the order is changed. According
to Piaget, she is experimenting not on the pebbles, but on her own actions
of ordering and counting. This situation differs from simple physical
experience because: (i) the actions 'enrich' the object with properties which
it did not have before, namely order and number; and (ii) the actions
performed by the subject are part of a wider activity pattern such that there
is always a tendency to introduce order into our movements, for instance.
Knowledge is 'reflectively abstracted' from these actions so that the actions
become internalized operations which describe the relationships, or coordinations, between the activities of counting and ordering. This is the beginning
of a process of successive reflective abstractions which results in fully
equilibrated, reversible, operational thought. I will discuss this process in
greater detail in Chapter 4.

In this account Piaget begins to explain the psychological possibility of
pure mathematics. As actions are transformed into operations, so these can
be carried out symbolically without reference to the original objects on which
the actions were carried out. And further:

Reflective abstraction starting from actions does not imply an empiricist interpretation in the psychologist's sense of the term, for the actions in question are not the particular actions of individual (or psychological) subjects: they are the most general coordinations of every system of actions, thus expressing what is common to all subjects, and therefore referring to the universal or epistemic subject and not the individual one. From the beginning, mathematics thus seems to be regulated by internal laws and to escape the arbitrariness of individual wills . . ..
And at each new stage the necessity of integrating, whilst going beyond, the results of the earlier constructions explains the fact that the successive constructions obey directional laws, not bec. use everything is given in advance, but because the need for integration itself involves a continuity which is only perceived retrospectively, but which nonetheless imposes itself.

(1966: 238)

Thus pure logic and mathematics become totally independent of the physical properties of objects, but they agree with experience and physical reality because of their origins in the cognitive, and consequently biological, organization of the subject:

as human activity is that of an organism which forms a part of the physical universe, it is understandable that these unlimited operational combinations should so often anticipate experience and that there is agreement, when they meet, between the properties of the object and the operations of the subject.

(1972b: 51)

The relationship between the development of thought and the development of mathematics

According to Piaget, the relationship between the development of thought and the development of mathematics has two aspects. The first aspect is that historical development in mathematics inverts the order of construction of thought; thus Cantor's use of one–one correspondence to develop the theory of transfinite numbers was a relatively late development in the history of mathematics because the act of putting objects into one–one correspondence occurs early in a child's development. This early occurrence means that the notion of one–one correspondence, which, according to Piaget, Cantor 'found in his own thinking' (1966: 242), was more difficult to unearth and bring into conscious awareness. Likewise, Piaget claims that topology is a more recent topic for mathematical study because spatial awareness is yet more primitive than one–one correspondence. A

22

related point is that formalization in mathematics, according to Piaget, inverts and refers to the construction of thought in the search for axioms and rules of demonstration (1966: 254).

The second aspect of the relationship between the development of thought and the development of mathematics is that progress in mathematics is an extension of the reflective abstraction of ordinary thought. Piaget claims that:

> from the genetic point of view formalization may well be regarded as an extension of the process of reflective abstraction already present in the development of thought. But because of the increased specialization and generalization which formalization possesses, it exhibits a freedom and a richness of combinatorial possibilities which largely transcend the bounds of natural thought.

> (1972a: 64)

There are two specific questions to be asked about the relationship which, in Piaget's view, can be answered by genetic analysis. These are: (i) 'What is formalized by logic?'; and (ii) 'Why are there limits to formalization in Gödel's sense?'

What is formalized by logic?

For Piaget, logic is the formalization of natural operational structures. This is the only remaining possibility, he says, after we have discarded the view that it is the axiomatization of our knowledge of objects (Piaget argues that physical objects exist in time and always change; thus the rules of logic must refer to *actions* carried out on objects); and the view that it is a simple syntax (Piaget argues that in such a case logic would no longer be a form of knowledge 'in the proper sense of the word' (1972a: 65) but a pure form concerned only with tautological relationships). The genetic data support the view that intelligence is prior to language and that pre-verbal intelligence shows a logical structure to do with coordination of schemes of actions.

Logic, then, formalizes natural thought, not in the sense of the 'conscious thought of the subject, with his intuitions and his experiences of self-evidence; for these vary during the course . . . of development' (1972a: 66) but rather in the sense that it axiomatizes the operations of thought. Thus logic

> expresses operations in the form of abstractions (classes, relations or propositions) which it manipulates in a purely deductive way, or in other words an axiomatic way, putting them into symbolic form the better to detach them from their mental context and to combine them rigorously. It is nonetheless true that the same operations are involved and that to every logical relationship there corresponds a real mental operation. Conversely, all equilibrated mental operations . . . can be expressed in the form of a logical relationship.

> (1972b: 79)

Piaget observes from the genetic data possibilities for formalizing structures both at the concrete-operational level where 'groupings' appear and at the fully operational level where structures corresponding to mathematical 'groups' appear. I will describe these in greater detail in the next chapter.

The limits of formalization

Piaget's second question regarding logic concerns the limits of formalization as shown by Gödel's theorem, that is, that the consistency and completeness of a system can only be demonstrated by a logic more powerful than that of the system itself. This limitation finds an analogy in Piaget's account of genetic development. Essential to his account is the assertion that, for the individual, successive constructions are qualitatively different in the sense of being more abstract and embodying a stronger logic in comparison to their predecessors. Thus Piaget writes:

> If the abstraction from actions and operations is necessarily reflective, that is to say, presupposes the reconstructions of the operational elements abstracted to form new operations, then the differentiation of a system through the analysis of its initial implications leads to a new, more abstract system. This system, because it is more abstract in this 'reflective' sense, is situated on another plane of construction and constitutes psychologically a new form of thought, subordinating and integrating the lower form, but sometimes contradicting the initial intuitions.
>
> (1966: 243)

One consequence of this process is the overlapping of form and content, so that broadly speaking the forms of one level ('forms' here meaning the structures) constitute the content of the next. So the structures of the concrete-operational level constitute at one and the same time the forms of the sensorimotor schemes and the content of the hypothetico-deductive operational structures.

The overlapping of form and content means that form cannot achieve a complete autonomy. Piaget explains:

> one sees that a form remains necessarily limited, that is, unable to guarantee its own consistency without being integrated in a more comprehensive form, since its very existence remains subordinated to the whole of the construction of which it forms a particular aspect.
>
> (1972a: 68)

So, for example, at the concrete-operational level there exist the two 'groupings' of classes and relations. While there exist implicit relations between these two, Piaget says, as long as the subsequent operational structure is not formed,

24

it is not possible to combine these two sets of 'groupings' of classes and relations in a unique formal system with coordination of inversions and reciprocities: their formalization thus remains incomplete so long as they are not integrated in a 'stronger' structure.

(1972a: 68)

The formalization of thought

Piaget also claims certain practical and theoretical advantages in a formalization of ordinary thought. He begins by acknowledging two possible objections against such a programme. One is that it is impossible since natural thought lacks the rigour necessary for axiomatization, and the other is that natural thought has specific properties precisely because it is not formalized, and which it would lose if transformed to a formal state. Piaget's counter-argument consists in saying that both of these objections assume that the only models of formalization possible are those formal systems already in existence which were constructed in the search for a foundation for mathematics rather than as models for natural thought. Now, Piaget says:

> Our aim is very different and is subject to neither of the preceding objections: it is simply to determine precisely the specific characteristics of some one or other structure of actual thought in its development, as well as its difference with perfected logics. If we suppose, for example, that the elementary 'groupings' . . . of classes and relations play an important part in development, and further . . . are at the starting point of the construction of natural numbers, it may then be interesting to formalise the structure of a 'grouping', not in order to assimilate it to a Boolean algebra, to a lattice etc., from which, precisely, it differs, but merely to bring out its specific limitations in a form which the logician can understand (even if it is of no use to him) and which it is useful for the genetician to know.

(1966: 256)

It is difficult to see that this statement answers either of the two criticisms. Piaget appears here to be merely softening his position (a formalization of thought would be 'interesting') in order to avoid having to defend it. Certainly, Piaget's thoughts on the connection between logic and psychology are much more complex than this: he goes on to say that, for the purposes of genetic analysis, formalization allows the exact isolation of lacunae at various levels, and thus what is necessary for development from child to adult. What is puzzling about this discussion of Piaget's is the apparent contradiction between: (i) the important idea that successive levels of development show different strengths of logic and thus imply a *qualitative* difference between levels; and (ii) the discussion here which indicates that different levels of development can be formalized accurately by using the same basic formal logic for all levels with the appropriate restricting postulates built in, thus resulting in what could be called a *quantitative* difference between levels. It would seem that these built-in 'lacunae' are an *ad hoc* device that cannot faithfully represent the quality of a particular level of development because the level in question is then being described in terms of a higher logic than that of its own system. While by Piaget's own argument it is unavoidable that the forms of one level constitute the content of the next, this result also runs counter to his very important points about growth, since it is the qualitative novelty of successive stages of development that ensures the lawful progress of mathematical thought in a particular direction, while at the same time preventing the prediction of this development before it happens. Piaget's use of a single logic plus lacunae to describe different levels of development means that his formalization must have a restricted use and value because it would not be accurate to use it as a description of the quality of thought at any particular level of development. It simply describes a final formalization and retrospectively states what is 'missing' from it, and thus fails to describe a level in its own terms. To put this another way, in Piaget's account one level is described from the point of view of another, higher, level. Piaget says that formalization can be used to describe what is missing at a certain level; this may be true but since it does not convey the quality of the level of thought in its own terms, it cannot then indicate why and how it develops from that level to the next, because it cannot accurately describe the starting point of development and the elements which lead to change. I will examine the consequences of this in Chapter 5.

Finally, Piaget considers that a formalization of ordinary thought brings together the genetic and axiomatic methods on an epistemological plane, by making clear the relations between 'elementary axiomatics' (identification of 'the axioms which are necessary and sufficient to deduce a system formally' (1966: 251)) and 'elementary genetics' (identification of 'the initial structures as well as the actions or operations which have made possible the transition from these structures to those whose development we must explain' (1966: 251)). The development of the number concept is an example for Piaget of 'the convergence between genetic and axiomatic investigations' (1966: 259) whose value is to 'show how certain general results of formal analysis are psychologically explicable if based on what we know of the subject's activities' (1966: 259). To go back to the foundations, where logicians have argued about whether number should be reduced to the logic of classes or the logic of relations, Piaget sees his task as establishing

whether the concept of an integer, as it is elaborated by 'effective' thought . . . can verify either of these two 'proposed' solutions. No doubt the objection will be raised that this concept of 'natural' number is not that of the mathematical sciences, which means that, even if the mind 'really' proceeds in a given manner, formalized theories can provide the foundation for the concept of number in their own way. Once again, however, it is clear that the idea of a concept of number preceded the formation of a scientific arithmetic, and that, if there exists an elementary intuition of the concept of number or a constitutional liaison between the concepts of number and class or logical relations, it is on pre-scientific grounds that the matter would initially have to be verified.

(1972b: 27–8)

Piaget's genetic epistemology is highly influential on his theory of number development in the child, and this theory in turn dominates the field in terms of the conceptualization of knowing number in psychological enquiry. In this chapter my purpose has been to set Piaget's theory in context with a view to clarifying his reasons for asking particular questions and giving particular kinds of answers to them. The next chapter describes the development of Piaget's theory of the number concept within this context, and raises some questions regarding the adequacy of the theory as a psychological account.

3

The child's conception of number

Piaget's criticisms of intuitionism and logicism

Genetic epistemology allows Piaget to use genetic data in order to make a judgement on the nature of the number concept, as part of the general programme suggested by the question, 'What makes mathematics psychologically possible?'. The foundations of arithmetic as examined by the intuitionist and logicist schools provide the starting point for Piaget's own account, which he presents as the result of a psychological investigation into the theories of number as the product of intuition and of number as the product of the logic of classes (Russell) or of relations (Peano). In the first section of this chapter I will discuss Piaget's (1966, 1968a, 1972a) criticisms of logicism and intuitionism, neither of which, he says, fit the empirical evidence. The main criticism of intuitionism lies in Piaget's assertion that perception of order (in this case the order of states of consciousness) is too complex to be directly intuited, being an active construction rather than passive reception. If number is not directly intuited, Piaget argues, then some other more primitive notion must be apprehended first. Piaget's treatment of logicism is initially directed towards settling, by means of genetic data, the question as to whether number should be reduced to the logic of classes or to the logic of relations. He also makes some theoretical criticisms of Russell's account of the natural numbers.

Intuitionism

The intuitionist tradition suggested that number is not reducible to logic, but rather is the result of the intuition of time. Piaget examines this possibility and argues against it from two viewpoints: (i) that the psychology of time is more complex than the intuitionist scheme allows; and (ii) that there can be no 'pure experience' of temporal succession as derived from objects themselves; instead order is introduced through our own actions.

Piaget (1966: 210) analyses four stages of the knowledge of time. First, sensorimotor time is sensitive to the order of succession of actions and the sensation of duration; second, perceptual time involves the perception of successions, simultaneities and duration; third, 'time lived through' is a semi-structured sense of time that is longer or shorter according to what one is doing which, while it is not purely perceptual, is not operationally structured either; and fourth, fully operational time is fully structured and synthesizes both serial operations and the overlapping of durations to generate a full understanding of time. This analysis demonstrates for Piaget the difficulty in the idea of intuited time: if intuition is characterized as immediate knowledge, then stage four time does not fit such a description, being an operational (and therefore complex) notion which takes some time to develop.

With particular respect to the reduction of number to the empirical intuition of time, Piaget argues as before that order is imposed by us, and is not inherent in objects themselves. Thus it is impossible for there to be an experience or intuition of order derived from objects or even states of consciousness alone. The perception of succession must be the perception of the succession of our own actions in, say, looking from A to B. Thus there is an intermediary between objects or events and our perception of order. In so far as intuition, by definition, is immediate knowledge, then this perception of order is not intuitive. Furthermore, Piaget says, the difficulties are compounded by the importance given to memory by the notion of intuition of time from states of consciousness. If these are experienced in immediate continuity so that they are comparable in pairs as they are lived through, then in order to perceive states of consciousness, memory is necessary. But, Piaget claims, the same argument as before applies: memory is not a question of merely noting order in the world, but it is a question of 'actively reconstituting' it. So 'order is known, not by abstraction from objects . . . but by abstraction from actions which order these objects' (1966: 213).

Parallelism in development between number, classes, and seriation

If number is not the product of immediate knowledge, then what is its basis? Logicians such as Russell and Peano have attempted to reduce number to logical entities, either classes or relations. Piaget's next step is to examine the developmental data in order to clarify the relationship between numbers and classes or relations.

First, he sets a minimum criterion for the acquisition of number since he is not satisfied with the simple ability to count verbally as an indication that a person knows number. Essentially, this minimum criterion is the ability to conserve number, operationally defined as the recognition that two rows of objects initially in perceptual one–one correspondence remain equal in number even when one of the rows is lengthened or shortened so that the perceptual correspondence is disturbed. Before this stage is reached (about 7–8 years), there are two intermediate stages. The first is characterized by the fact that the child does not use one–one correspondence as an indicator of equivalence. Thus, he may judge a row of six elements to be equal to a row which has the same length but which contains a different number of elements. At the second stage the child acknowledges one–one correspondence as a criterion of equivalence, but only if this is perceptual; rearrangement of one row leads to non-conservation.

Piaget observes a correspondence between these three stages and the three stages in the development of class inclusion. At the first stage, when given a collection of objects to be classified, the child makes 'figural collections', that is, collections of elements are made not only on the basis of resemblances and differences, but also according to the relations between things – for instance a table and what goes on it. In addition the elements themselves are arrayed in spatial configurations such as rows or squares. At the second stage, the child no longer makes figural collections, but instead constructs collections which overlap. However, this overlapping is the result of trial and error and is not anticipated; the child does not fully understand the quantifiers 'all' and 'some' and so does not see the necessity that if B = A + A' then A < B and A' < B. The third stage sees the completion of this understanding of class inclusion.

Again, these three stages correspond to the development of seriation or asymmetrical transitive relations. At the first stage, a child given, say, the task of ordering ten elements in terms of size, only succeeds in creating pairs of small versus large, or small ordered groups. At stage two, the child successfully orders all ten elements after some trial and error, while at stage three he orders systematically according to the rule of successively picking out the smallest of the group elements.

The evidence for a parallelism between number, classes, and seriation is strengthened by some further observations of children's behaviour. For instance, Piaget notes that a child may act as though he considers that five elements taken from a collection of ten are less numerous than five similar elements taken from a collection of thirty or fifty. This corresponds to the first stage of class inclusion, and indicates, according to Piaget, an incapacity to separate extension (the things denoted by a term) and intension (the essential qualities or definition of a term). Another example that Piaget cites which connects seriation and number is the observation that when a child (prior to stage three) is given a collection of, say, thirty elements, he does not appear to realize that in having taken one element away at a time he must have passed through a stage where there were fifteen elements. So, to return to the question of intuition, Piaget says that the protracted period of development described here firmly discards any argument for an intuition of n + 1 'in the sense of a clear awareness of iteration and recursive processes' (1966: 262):

We may thus conclude that in the domain of natural thought as from the standpoint of formalisation, the construction of number proceeds from the logical elements of classes or relations and is not an independent elaboration based on intuitions which are at once primitive and *sui generis*.

(1966: 263)

Logicism

Piaget's next step in this analysis of the psychological basis of number is to examine whether number should be reduced to the logic of classes or relations. To this end he examines Russell's (1919) characterization of natural numbers. Russell defines number informally as:

> what is characteristic of numbers . . . . A trio of men, for example, is an instance of the number three, and the number three is an instance of number; but the trio is not an instance of number . . . . A particular number is not identical with any collection of terms having that number: the number three is not identical with the trio consisting of Brown, Jones and Robinson. The number three is something which all trios have in common, and which distinguishes them from other collections.
>
> (1919: 11–12)

Russell's definition of number involves the idea of bringing together collections that have a given number of terms. Thus we can bring together couples, trios, and so on in bundles. Since each bundle is a class whose members are collections, each can be called a 'class of classes'. So the bundle of all couples is a class of classes where each couple is a class with two members and the whole bundle is a class with an infinite number of members, each of which is a class of two members (Russell 1919: 14).

When it is necessary to decide whether or not two collections belong to the same bundle, a method of deciding must be used which does not involve number or counting, since this is what is being defined. The method is that of using one–one correspondence, and two classes can be said to be similar when a one–one relation correlates the terms of one class with the terms of the other class.

The notion of similarity is used to decide when two collections must belong to the same bundle, and, given any collection, the bundle it is to belong to is defined as being the class of all those collections that are similar to it. Having set up this definition, Russell then defines number itself:

> We naturally think that the class of couples (for example) is something different from the number 2. But there is no doubt about the class of couples: it is indubitable and not difficult to define, whereas the number 2, in any other sense, is a metaphysical entity about which we can never feel sure that it exists or that we have tracked it down. It is therefore more prudent to content ourselves with the class of couples, which we are sure of, than to hunt for a problematical number 2 which must always remain elusive. Accordingly we set up the following definition: *The number of a class is the class of all those classes that are similar to it.*
>
> (1919: 18)

Thus the number two is in fact the class of all couples. According to Russell, this definition generates all the requisite properties of number, and also 'secures definiteness and indubitableness' (1919: 18). More generally, numbers can be defined as any one of the bundles which classes are collected into by virtue of their similarity, so that a number is any collection which is the number of one of its members. Finally '*A number is anything which is the number of some class*' (1919: 19).

Piaget's criticism of this is that there is a problem in Russell's definition when seen from the psychological point of view, and possibly from that of logic too. This is that there are two types of one–one correspondence, and that Russell confounds them. The two types of correspondence are: (i) a qualified correspondence, which means that in two corresponding sets, each object in the first set corresponds to one in the second set which shares the same quality of, say, being a square, the same colour, etc.; and (ii) a generalized correspondence, where objects are put into one–one correspondence regardless of their qualities. Now, according to Piaget, in order to constitute the second type of correspondence the individual objects must become arithmetical unities because they are stripped of their qualities. If they are no longer circles, squares, red counters, and so on, then they are simply units and therefore have a numerical quality. So, Piaget argues, Russell (1919) and Russell and Whitehead (1910–13) have 'smuggled in' the idea of number into their definition because they do not use qualified correspondence which is based on *things*, but instead use the generalized correspondence which can only operate with numerical units:

> Now it is very clear that Russell and Whitehead have not used the qualified one–one correspondence that is used in classification (i.e., qualitative one–one correspondence). They have used the correspondence in which the elements become unities. They are, therefore, not basing number only on classification operations as they intend. They have, in fact, got themselves into a vicious circle, because they are attempting to build the notion of number on the basis of

32

one–one correspondence, but in order to establish this correspondence they have been obliged to call upon an arithmetic unity, that is, to introduce a notion of a non-qualified element and numerical unity in order to carry out the one–one correspondence. In order to construct numbers from classes, they have introduced numbers into classes.

(1968a: 27)

If the idea of number is not introduced into generalized one–one correspondence, then since elements are no longer distinguished by their qualities, it becomes impossible to tell them apart, Piaget argues. What should be the basis of a numerical series for Russell and Whitehead collapses into the tautology $A + A = A$, unless some other way of distinguishing the elements apart from their qualities can be put into operation. For Piaget the introduction of an order relationship is the only way of distinguishing elements which would otherwise be considered as identical; and indeed, he argues, Russell and Whitehead's account does in fact make implicit use of the notion of order as well as that of class:

If we examine closely classical reductions . . . we discover in each of them a twofold appeal to classes and to the relation ($<$) . . . [Russell and Whitehead's] solution is to contrive that two number-classes A and B, which represent the terms of a sum, have never the same members; and to do this, they simply introduce differences of order.

(1966: 270–1)

The means by which Russell and Whitehead introduce order is in the form of ordered pairs. The question here is: are ordered pairs needed to give a full description of the system of natural numbers, and in particular are they necessary for the definition of addition? Piaget says they are. To quote him almost verbatim (1966: 271), he says: let for example, $A = (a,b,c)$ and $B = (a,d)$. The operation of $A \cup B$ (union) combines the elements of A and B to give $C = (a,b,c,d)$. Now the operation $A + B$ comes to that of $A' \cup B'$ where $A' = [(a,0), (b,0), (c,0)]$ and $B' = [(0,a), (0,d)]$. A and A' and B and B' have the same force so that

$$(A + B) = (A' \cup B') = [(a,0), (b,0), (c,0), (0,a), (0,d)]$$

So, Piaget argues, in order to avoid the tautology $A + A = A$, only the introduction of the order $A \rightarrow A$ in the form of distinguishing $(a,0)$ from $(0,a)$ will give the requisite distinction between the elements with the result $A + A = 2A$. This, he says, is supported by the genetic data such that the method above is 'similar to the way in which natural thought brings in a reference to order to distinguish two elements which are different though equal' (1966: 271).

The child's conception of number

Piaget's synthesis of order and class as logical criticism

There are several points to be made against Piaget's criticisms. First, ordered pairs can themselves be reduced to a definition in terms of sets of sets (Barker 1964: 60); this means that Russell and Whitehead's thesis does not involve the notion of order as a primitive idea, since it is expressed in the form of sets. Second, in any case, the 'order' of ordered pairs is very different from the transitive asymmetrical order that Piaget argues is a necessary component of number as it is psychologically understood. According to Barker (1964: 60), an ordered pair 'consists of two things of any kind what ever, considered in a certain order. The things may be concrete or abstract, similar or dissimilar'. An ordered pair x;y is identical with another ordered pair z;w if x is identical with z and y is identical with w. By contrast, transitive-asymmetrical order relates individual items into pairs in terms of relations such as longer than, taller than, and so on. More formally, a relation '<' is a transitive-asymmetrical relation if, for any A, B, C, if A<B and B<C, then A<C, and if A<C then not C<A. An ordered pair x;y, however, need have no such relation between x and y, the only important factor being the order in which they are considered.

A third point against Piaget is that Russell and Whitehead's formalization is sufficient to express the principles of the natural numbers, and although the notion of ordered pairs must be used for the definition of addition it is not necessary for the definition of number. Additionally, Piaget's original argument concerning the inevitable tautology of generalized correspondence without order is incoherent. First, recall Piaget's argument as it is expressed in a different source:

> A first peculiarity of a numerical collection . . . as opposed to collections which are simply classifiable or serializable, is that it abstracts the qualities of individual terms, so that they all become equivalent. They could then still be arranged in the form of overlapping classes (I + I) (I + I + I) etc., but only on condition of their being distinguished one from the other, since otherwise one element could be counted twice or another overlooked. Now, once the differentiating qualities of the individuals I, I, I, etc. have been eliminated, they become indiscernible and, if one restricted oneself to the operations of the logic of qualitative classes, could only yield the tautology A + A = A and not the iteration I + I = II.
>
> (1972a: 38)

What is the force of this argument? The preceding discussion shows that Piaget does not intend his criticism to be a merely psychological or commonsense point. As the above quotation tends to indicate, rather, he means the argument to have logical force too. Consider, then, in what logical sense things become indistinguishable: Piaget gives the impression that when the qualities of objects are ignored, they lose all substance and roll into one like droplets of mercury. And, he says, if things are considered as units, then this is making use of the concept of number and considering them as instances of the number one, which is the very thing to be defined. So, in order to stop objects becoming indistinguishable, it is necessary to locate them spatially or temporally, if use of the concept of numbers is not allowed. The problem with this description is that, in the first place, the idea of stripping objects of their differentiating qualities does not mean they no longer exist as entities unless considered as arithmetic units; they can still be elements of a set, a notion which does not presuppose number. It is possible to consider objects as things without regard to either distinguishing qualities or numbers. While it is true that things are separate from each other in time or space or both, it does not seem to follow that in order to count these it is necessary to have a logical idea of spatial or temporal order. Piaget's claim that we need to be careful not to count things twice or overlook things seems to be something of a commonsense truism that bears no force in logical argument. As Russell points out:

What we do when we count (say) 10 objects is to show that the set of these objects is similar to the set of numbers 1 to 10 . . . . In counting, it is necessary to take the objects counted in a certain order, as first, second, third, etc., but order is not the essence of number: it is an irrelevant addition, an unnecessary complication from the logical point of view. The notion of similarity does not demand an order: for example, we saw that the number of husbands is the same as the number of wives, without having to establish an order of precedence among them.

(1919: 17)

So, as the case of infinite sets demonstrates, one–one correspondence does not rest on the ordering of terms, or even on knowing all the terms, but rather on the function which maps elements onto each other.

A last point to make is that, despite his efforts to bring a logical criticism of Russell's formalization, Piaget's argument has finally collapsed into psychologism: he refers to mere commonsense psychological observation for support, and fails to introduce convincingly the strong logical sense of transitive-asymmetrical relations which his psychological theory demands. The introduction of order relations only has force from a psychological point of view, and even this is weak. In the next section, I will discuss in greater detail Piaget's account of the psychological development of the number concept as a synthesis of order and class.

Piaget's synthesis of order and class

This section describes in detail Piaget's alternative account of number as a synthesis of order and class. This account is woven into his general theory of successive developmental stages characterized by increasingly elaborate logics or ways of thinking, and it describes quite precisely the logic of the concrete operational stage of development.

The growth of the number concept takes place during the pre-operational stage, which is characterized by figurative thinking where perception dominates, and the concrete operational stage, characterized by thinking which is reversible but limited to real situations. In the pre-operational stage, intelligence is characterized by its internalization of the action schemes of the sensorimotor phase so that, instead of merely acting on objects, the child acts in thought on *representations* of objects. However, figurative thinking leads to the inability in pre-operational children to do such things as conserve liquids, decentre themselves from their own perspective, and to seriate or classify objects without trial and error; pre-operational intelligence is marked by its lack of reversible internalized operations. Developing towards concrete-operational intelligence requires that internalized action schemes are coordinated (as illustrated in the last chapter in the example of the pebbles) to form a 'scheme of schemes'. While thought is now 'operative' – that is to say, it is reversible – this stage is limited in that it cannot deal with situations absent from the senses. It is at this stage that the concept of number is complete.

The ability to reason about theoretical possibilities marks the next and final (i.e., adult) stage of thought, called formal operational or hypothetico-deductive thought. This demands a comprehension of propositional logic, which involves the ability to form operations on operations. Problems of proportion involving relations rather than simple reversal of actions become solvable at this stage. This is possible because the two kinds of reversal of action, inversion (cancellation of an action by another such as pouring liquid back into a glass) and reciprocity (a second action follows in the opposite order to the first), become coordinated as part of the same group structure. According to Piaget, this coordination of the two kinds of reversal in fact uniquely occurs in the concrete-operational stage with respect to number, and thus singles out the development of the number concept as an achievement of note.

The structure of concrete-operational thought

Piaget begins his analysis of concrete-operational thought by introducing the notion of a 'grouping':

If we mean by 'operations' actions which are interiorised, reversible (in the sense of capable of being carried out in both directions) and co-ordinated into structured wholes; and by 'concrete' operations those occurring in the manipulation of objects, or in their representation accompanied by language, but not concerned solely with propositions or verbal statements . . . all the structures at the stage of concrete operations are reduced to a single model, which may be given the name of a 'grouping'.

(1966: 172)

Piaget describes a grouping further as 'a system lacking logical generality because of its multiple restrictions' (1966: 172); these restrictions are most clearly seen when the grouping is compared to a group, which is defined by Rotman (1977: 48) as a set T of transformations which has the properties of: (i) closure: if x, y are members of T then xy also belongs to T; (ii) associativity: if x, y, z, are members of T then x(yz) = (xy)z; (iii) unity: there is a unit transformation u in T having the property that ux = xu = x for all transformations x in T; and (iv) inverse: for any transformation x in T there is an inverse transformation y in T such that xy = yx = u.

Now, groupings differ from groups in two important ways. First, groupings have a step-by-step (or contiguous) composition, so that elements cannot be combined with complete freedom, as is the case with groups. Second, while groups are associative, the associativity of a grouping is restricted so that (A + A) − A is not equal to A + (A − A). This is due to the presence of the tautology A + A = A. These points are illustrated in the classification grouping: if A, B, C, etc. are overlapping classes with complementaries A', B', C', etc., then certain rules hold as follows:

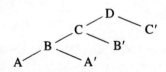

(i) A + A' = B; B + B' = C, and so on
(ii) B − A = A'; C − B = B', and so on
(iii) A + 0 = A
(iv) A + A = A (tautology)
(v) (A + A') + B' = A + (A' + B')
but (A + A) − A ≠ A + (A − A)
for A − A = 0 and A + 0 = A
A non-contiguous composition such as A + C' yields the complicated class D − B' − A'.

Now, Piaget says that:

These restrictions are very significant from the psychological viewpoint: combinations which are exclusively of a step by step form, express, in effect, a beginning of deductive power, not yet freed from concrete manipulations and only proceeding thus by means of contiguous overlapping without achieving a combinatorial system.

(1966: 174)

The other important grouping involved in the development of the number concept is that of seriations (transitive-asymmetrical relations). The relations in an additive grouping of asymmetrical-transitive relations can be symbolized as follows:

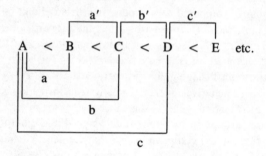

This gives the result:

$a + a' = b$; $b + b' = c$; $c + c' = d$, etc.

$b - a' = a$; $c - b' = b$, etc.

$a - a = 0$; $a' - a' = 0$

$a + a = a$

$(a + a') + b' = a + (a' + b')$ but $(a + a) - a \neq a + (a - a)$

for $a + a = a$ and $a - a = 0$ and $a + 0 = a$

Here again the characteristics of groupings as Piaget lists them can be seen, and the same restrictions are evident. These two groupings have their major difference in the types of reversibility that they employ. Inversion or negation, which is characteristic of the additive grouping of classes, takes the form $+ A - A = 0$. Thus if one begins with a class x, adds class A and then subtracts it, the effect is neither to add nor take away anything. Reciprocity is the reversibility characteristic of ordering structures, and is not a negation but a reversal of order. Instead of A<B, the result is B>A:

$(A<C) + (C>B) = (A<B)$ or $b + -a' = a$

$(A<B) + (B>A) = (A = A)$ or $a + -a = 0$

Here, reversibility as represented by reciprocity does not reduce to an annulment as does inversion, but rather 'the relation designated by 0 is not the suppression of a relation but the suppression of a difference, which leads to a relation of equivalence (A = A)' (Piaget 1966: 178).

The relationship that Piaget observes between the two kinds of reversibility is a significant one. At the concrete-operational stage they are independent, and they remain so until the beginning of hypothetico-deductive operations at age 12 to 15 years, so that it is not possible, before this age, to pass from one type of reversibility to the other *within the same system*. With the advent of the number concept and the abstraction of qualities, however, a particular convergence occurs where the two groupings are synthesized into the same system. It is important to distinguish, though, between the synthesis that occurs in this special case and the true combination of reciprocity and inversion that occurs at the hypothetico-deductive stage. Piaget writes that at the latter level

> we see the emergence of a new structure which results from a second form of connection between structures involving inversions and those involving reciprocities. We shall here no longer speak of 'synthesis' as in the case of the formation of number, but of a 'combination' in the sense that the new structure will involve, on the one hand, transformations N in the form of inversion and, on the other, transformations R in the form of reciprocity, both remaining distinct but capable of being combined together, as opposed to what occurs at [the concrete-operational stage].
>
> (1966: 180)

Number as the synthesis of two groupings

Returning to the question of the development of the number concept, recall Piaget's original question, which is: How is a transition from class to number effected, or, how is it that elements from which the qualities have been abstracted and which are thus made equivalent can still be distinguished?

> Let the singular classes A_1, A_2, etc., be . . . distinguished by their qualities: once the latter are eliminated, how are we to explain why the subject does not end up with the tautology A + A = A, since without distinctive qualities we have A_1 = A; A_2 = A; etc., and that, nevertheless, he arrives at the iteration A + A = 2A, from the fact that one of these A's is distinguished from the other in spite of the absence of distinctive qualities? In concrete terms, the question is reduced to the following, for example: if for a collection of counters of individually different colours etc., we substitute a collection of counters of the same dimensions and colours, how will the subject distinguish between them (in the operation of making them 'correspond in any way whatever')?
>
> (Piaget 1966: 265)

39

The child's conception of number

As is clear from the preceding discussion, Piaget's answer is to synthesize the two groupings – classifications and seriations. Initially, the two groupings are independent in that they cannot both be applied to the same elements, as long as these are qualified, for they must either be considered according to their partial equivalences (classification) or their orderable differences (seriation) (1966: 260). However, Piaget argues, once the qualities are abstracted (but note the problem with this circular explanation), certain necessary consequences follow 'which we maintain are sufficient to explain the formation of natural numbers' (1966: 260). The first of these consequences is that the groupings necessarily merge into a single system; one cannot function without the other. As I have already noted, the structure of Piaget's argument demands a certain instantaneous and presupposing quality of this combination: the maintenance of the class structure assumes the introduction of the order structure, while at the same time the exploitation of the order structure presupposes the class structure. This is because to order equivalent elements we need to consider the first element as preceded by the null class, the second as preceded by the class (A), the third by the class (A + A), and so on.

The second consequence is that the limitations of groupings are suppressed. Recall that these are: (i) the restriction of associativity to the non-tautological operations; and (ii) contiguous composition. The synthesis of the two groupings, however, means that: (i) the elements are all substitutable for each other so that the order A → A → A is conserved however the elements are permuted ('vicariant order') with the result that the singular classes of the groupings of classes are equivalent but not tautological, being distinguished by their vicariant order; and (ii) the classes A + A = B; B + A = C and so on

> remain the same if the A's are permuted, which means that any A can be combined with any other in a class of category B, without taking further account of contiguity or step by step order (generalised overlappings based on vicariant order).
>
> (Piaget 1966: 267)

Thus, (I) < (I + I) < (I + I + I), etc.

The third and final consequence is that the series of unit elements A → A → A which combine as A + A = B, B + A = C, etc., has the characteristics of the number series. The newly synthesized system has the formal properties of the natural numbers.

Piaget (1966: 268ff.) reports the formalization of number as a synthesis of class and order by his co-worker Grize (1960). Grize first formalizes the structure of a 'grouping' by introducing certain restrictive postulates (Piaget 1966: 173) which serve to generate the limited combinatorial range of the grouping structure. To formalize the synthesis of number (Piaget 1966: 268)

Grize modifies these postulates so that certain restrictions, in particular that which leads to tautology, are eliminated. Grize is able to derive theorems and metatheorems, including Peano's five axioms and a definition of addition. So, Piaget finally says that:

> Grize's formalisation thus shows that the natural process observed in the child's construction of number may correspond to a formal construction . . . . The principal difference between the usual formalisation, and the process of natural construction, which we think we have brought to light and which has been formalised by Grize, is that in the latter we explicitly deal with a synthesis between the overlappings of class and seriation, whereas in formalisations of the apparently deductive type as opposed to the synthetic or dialectical type, we appeal both to classes and to asymmetrical relations; or else, like Russell and Whitehead, we at first refer to only one of these structures, reintroducing the other almost surreptitiously later, in the guise of an expository device or construction. Such observations are therefore as much in favour of a convergence between natural processes and formalisation as Grize's formal construction, which is explicitly intended to correspond to the observed genetic scheme.
>
> (1966: 270–2)

In this chapter I have covered the background to Piaget's theory of number development and the theory itself in some detail. Piaget's theory is the most highly developed and influential of psychological theories of number development; as Chapters 2 and 3 have shown, his account of the development of the number concept is just one part of a wider theory which encompasses many aspects of knowledge. Piaget consciously uses and acknowledges the philosophical influences on him, and, more importantly, claims that his theory is largely an attempt to solve a philosophical problem – that of what makes mathematics true. This fact suggests that, if problems of epistemology such as concern Piaget were to be ignored, then a psychological theory might arrive at a completely different characterization of knowing about number. This is particularly relevant with respect to the many later theories of number development inspired by criticism of Piaget, which are psychological theories first and foremost and have no claim to being part of epistemological study, but yet make the same assumptions as Piaget about what knowing number must involve.[1]

The details of the structure of operational thought and the logical structure of knowing about number described in the second section of this chapter illustrate Piaget's serious intention, in setting out to formalize thought, of contributing to the epistemological debate. Here it can be clearly seen that Piaget uses a single system of logic with built-in restricting postulates to describe more primitive levels of thought. I have already made some initial criticisms of Piaget's account of knowing in Chapter 2: the assumption that different stages of development are characterized by different qualities or strengths of logic is essential to his theory, but there is a question as to whether this description of thought in terms of formal logic is adequate from a psychological point of view. While it is not clear that a description of a 5-year-old's knowledge in a 5-year-old's terms would be either possible or useful, a characterization of a child's knowledge in terms of formalized adult knowledge with 'restricting postulates' is not necessarily useful either, if the aim is to account for how the child gets to this point or carries on from it. One of my aims in this book is to argue for an account of knowing about number which includes an analysis of how far a child can be said to have entered into the social practices involved in using numbers; such an account would not involve formal descriptions of either children's or adults' knowledge.

In this chapter I have discussed further Piaget's philosophical influences and re-emphasized his aim of solving a problem of epistemology. Piaget attempts to criticize Russell and Whitehead's account of number by appeal to logical argument and also to psychological fact. Neither of these are convincing, in particular Piaget's criticism of Russell and Whitehead's logic, which at times collapses into psychologism. From the point of view of psychology, Piaget also fails to argue with any force for his importation of logic into psychology; specifically his claim that objects stripped of their qualities become indistinguishable, even if true, does not necessarily lead to the need for a logical idea of spatial or temporal order. In giving an account of Piaget's theory of number as a synthesis of order and class, I have also introduced my other main purpose, which is to give an alternative account of how knowledge about numbers is acquired. An initial criticism made in this section is that the claimed synthesis of the two groupings into one system has to be instantaneous in order to have any logical coherence, and there is consequently a suggestion of circularity in Piaget's account. This synthesis is the point of growth in this particular stage of the development of the number concept, but Piaget's account is lacking in any detailed description of what is happening at this point. This is a major feature of Piaget's theory, and although he attempts to describe the mechanisms of change from one strength of logic to the next, this aspect of his account is the least well-developed.

One reason why this might be so is that Piaget's initial characterization of knowing number proscribes any adequate description of development, simply because it is difficult to describe a transition from weaker to stronger logics without a simplistic falling away of restrictive postulates as in the synthesis of the two groupings described in the second section. The next two chapters consider the problem of development in more detail: Chapter 4 describes the problem of growth as Piaget sees it and his proposed solution, and Chapter 5 concentrates on a detailed criticism and begins to suggest some alternatives.

4

Piaget's account of the growth of understanding

Piaget's criticisms of empiricism and rationalism

In Chapter 2 I described how Piaget's account of the apparent necessity of mathematical statements involves an assertion of parallelism between the structures of thought and the structures of mathematics or logic. In Piaget's view, elementary logico-mathematical experience can be characterized in three ways: as the empirical observation of relations between objects in external physical experience; or as the domain of rational internal psychological experience; or again, as it is for Piaget, as a mixture of these two, a middle way between mathematics as discovery and mathematics as invention, in which it is human action that provides the raw material for mathematical constructions, not the objects which are acted on.

This chapter examines Piaget's criticisms of empiricism and rationalism and describes his own proposed middle way in accounting for the growth of understanding. As I noted in the last chapter, the way in which Piaget (together with later theorists) deals with the problem of describing development is of major importance, and is linked to the question of what constitutes an adequate account of knowing about number. In the following chapter I will argue that some of Piaget's criticisms of other accounts of development may in fact be applicable to his own theory, and that this is an outcome which is inevitable given the nature of Piaget's characterization of knowing about number.

Piaget's criticisms of empiricism and rationalism are that empiricism is implausible because of the passivity that it ascribes to the knower, while rationalism as embodied in Chomsky's (1957, 1965) theory of language acquisition for instance fails to explain: (i) the psychological evidence that shows a growth of reason (as opposed to the existence of innate principles); and (ii) how it is that innate structures of reason should coincide with the laws of nature.

The criticism of empiricism

For Piaget, an empiricist view of learning is inadequate because of the passivity that it ascribes to the knower. Empiricism, he says, gives a picture of cognition as a 'submission' to reality, where the mind's only function is to produce a mirror image of the world. Piaget also objects to the atomism of empiricism, arguing that its reduction of knowledge to elementary units cannot account for the creativity of learning. He hypothesizes instead that:

> Our various forms of knowledge originate neither in sensation nor in perception alone, but in the whole of behaviour, in which perception merely performs the function of signalling. The attribute of intelligence is not, in fact, to contemplate but to 'transform', and its mechanism is essentially an operational one. The operations consist of actions which are internalized and coordinated into group structures (reversible, and so on) and if we wish to give an account of the operational aspect of human intelligence, it is therefore action itself and not perception alone which provides an appropriate point of departure.
>
> (1972b: 47–8)

Piaget states this hypothesis strongly when he says that a person cannot understand the properties of an object except through action and transformation. Such transformation takes two modes: one is the empirical investigation of an object's nature as manifested by its positions, movements, and properties; the second is the addition of new properties or relations. This contrast has been drawn before in Chapter 2, where the observation of the invariance of a number of pebbles however they are counted is an instance of the second type of transformation which is logico-mathematical activity; whereas the observation of the weight of the pebbles is of the first type of transformation, and is empirical activity. A possible objection to this is that action itself is only known to us through proprioceptive perceptions. Piaget's counterargument is as follows:

> The relevance of this as far as knowledge is concerned is not the sequence of such actions considered in isolation: it is the 'schema' of these actions, that is to say that which is general in them and can be transposed from one situation to another (for example a schema of order or a schema of collections etc.). The schema is not drawn from perception, proprioceptive or otherwise. It is the direct result of the generalization of actions themselves, and not of the perception of them and it is not in the least perceptible as such.
>
> (1972b: 48)

Thus knowledge is constituted by the observation of patterns of behaviour, not the behaviour itself.

Piaget and the growth of understanding

Piaget's distinction between logico-mathematical knowledge and empirical knowledge is that the first can be said to be abstracted from ways of acting on objects, while the second is knowledge of objects which results from abstraction from their physical properties and so on. Hence the importance of action in logico-mathematical knowledge, to the extent that logico-mathematical knowledge and experience can become independent of objects themselves:

> This is why there exists, beyond a certain level, a pure form of logic and of mathematics to which experience becomes irrelevant. This is also why these pure forms of logic and mathematics are capable of overtaking experience indefinitely, since they are not limited by the physical properties of the object.
>
> (1972b: 51)

In contrast, empirical knowledge has more to do with perception, since it proceeds by abstraction based on the properties of the object as such. Thus, for Piaget, simple abstraction is sufficient for the gaining of knowledge modifiable by experience, but must be rejected as the means by which we gain what for him is unfalsifiable mathematical knowledge. However, Piaget argues, we should not make the mistake of thinking that perception is the sole origin of empirical knowledge, for the properties of objects cannot be discovered without the logico-mathematical frameworks which are necessary for perception:

> If there exists a pure form of logico-mathematical knowledge – pure insofar as it is detached from all experience – there does not exist reciprocally any experimental knowledge which can be qualified by the term 'pure' in the sense of being detached from all logico-mathematical frameworks, consisting in classifications, ordering, correspondences, functions, and so on. The act of perceptual scanning itself presupposes . . . the intervention of frameworks of this kind or of their rough outlines, more or less differentiated.
>
> (1972b: 52)

Thus, for Piaget, even at the beginning of life, perceptual data are organized according to the intervention of sensorimotor schemes. I will examine how this can be so in the next section. So, Piaget concludes,

> if action and intelligence transform perception in return, and if this latter, so far from being autonomous, is more and more narrowly structured by preoperational and operational schematism, the hypothesis of the sensory origin of knowledge, is then, to be considered not only incomplete, . . . but even false in the area of perception itself. False,

that is, to the extent that perception as such cannot be reduced to a scanning of sensory data, but consists of a process of organisation foreshadowing intelligence and increasingly influenced by its progress.

(1972b: 57)

The criticism of rationalism

The classical denial of empiricism is rationalism: reason rather than experience is the means by which we know the world. Here, Piaget is dissatisfied with what he calls a theory of structure without genesis, and he directs his argument towards rationalism as it appears in Chomsky's theory of language acquisition. While he agrees with Chomsky's (1959) criticisms of behaviourist and associationist models of language learning, Piaget argues that the theory presents the problem of explaining biologically how an innate language acquisition device is possible, and that, more generally, rationalism always presents the problem as to the origins of innate principles of reason: the genetic evidence of the growth of logical thinking in children refutes the strong form of rationalism which states that the principles of reason are innate. Piaget's main criticism of Chomsky's theory, though, is that it entails that the schemes that the child applies to the world in the early stages of life are not modified over time but on the contrary remain the same in later life. A related point is Piaget's criticism of rationalism for its failure to explain how the innate structures of reason coincide with the necessary physical laws of the world; genetic epistemology entails that the structures of knowledge achieve necessity at the end of their development rather than having it at the beginning.

Kant's (1933, 1977) theory of knowledge provides the traditional solution to the joint problems of empiricism and rationalism. His 'third way' entails that to be objective a judgement about the physical world must conform to synthetic *a priori* principles. While Piaget has much in common with Kant, he tends (perhaps wrongly, since Kant was not concerned with the problem of development but with that of how we can claim to know anything) to identify the Kantian *a priori* with innateness so that his criticisms of Kantianism coincide, in two works at least, with criticisms of rationalism or biological innatism. For instance:

It seems genetically clear that all construction elaborated by the subject presupposes antecedent internal conditions, and in this respect Kant was right. His *a priori* forms were, however, much too rich: he believed, for example, in the universal necessity of Euclidean space, whereas non-Euclidean geometries have reduced it to a particular case . . . So it seems that if we wish to arrive at an authentic *a priori*, we must progressively reduce the 'intension' of the initial structures until what remains qua antecedent necessity is reduced to a simple functioning.

(1972a: 91)

And elsewhere:

The operational constructivism suggested by genetic analysis is reduced neither to empiricism nor to apriorism, because we could not derive intelligence itself from objects . . . and because the subject does not possess frameworks which contain all reason in advance, but only a certain activity which allows him to construct operational structures. This construction is not arbitrary, for the individual subject is neither its origin nor does he seem to control it. The epistemic subject . . . is what all subjects have in common, since the general coordinations of actions involve a universal which is that of biological organisation itself.

(1966: 285)

Aside from this rejection of Kant's so-called innatism, Piaget is critical of his lack of a scientific method, in which respect he is critical of all epistemologies (1972b: 4–5). However, Piaget's theory is very much what would be expected of a Kantian system applied to developmental problems. Parallel to Kant's synthetic *a priori* principles are those of the process of genesis with structure which are necessary for an individual to acquire objective knowledge. As the preceding chapters have shown, Piaget's argument is that objectivity is the product of activity: objective thought is fully equilibrated thought, equilibrium having for Piaget a biological necessity; for Kant too, objectivity is a function of the human mind and is not independent of it. Piaget comments:

So we see that, without ending up with an apriorism in the traditional sense of a preformation of knowledge in forms which would contain it in advance, the contructivism to which genetic analysis has led us is related to apriorism in one sense, since each new construction derives its elements from a simpler construction previously carried out at a lower stage, and so on, this endless regression being due not to a fault in the system, but merely to the actual unknowns of biological morphogenesis.

(1966: 298)

Piaget's criticisms of empiricism, rationalism and Kantianism thus introduce his own particular synthesis of genesis and structure. The next section deals with this synthesis in greater detail on the psychological level, concentrating on the mechanisms of growth: assimilation, accommodation, equilibration, and reflective abstraction.

Growth as a synthesis of genesis and structure

Piaget's account of the development of the number concept, which is strongly influenced by his philosophical concerns, is in its turn highly influential upon other psychological theories, which include elements of Piaget's theory and approach in more or less explicit forms. His description and explanation of development is recognized as problematic by most theorists but, as later chapters will show, their alternative accounts do not depart in any substantial way from Piaget's general approach. It is my aim in this chapter and the next to point out the problems in Piaget's description of development, the main criticism being that he has little success in his attempt to improve on the abstraction model of empiricism through his synthesis of genesis and structure.

A successful account of this synthesis is of central importance to Piaget's theory since it would ensure an explanation of the dynamism of development which, as Piaget says, is missing from empiricist and rationalist accounts. Thus a large part of Piaget's work (1966, 1968b, 1972a, 1977, 1978a, 1978b, 1980) involves the elaboration of a description of development as a dynamic process which can explain how the child's thought progresses from sensorimotor intelligence to operational intelligence. Piaget's aim is to explain the growth of new knowledge as a qualitative rather than a quantitative event, without recourse to empiricist assumptions of an unstructured accumulation of perceptions, or rationalist notions of an automatic maturation of innate structures.

The dynamism in Piaget's account centres on the notion of equilibration: the driving force of human life, both cognitive and biological, is its progression towards more stable states of equilibrium between the self and the environment; human knowledge progresses towards a greater equilibrium, which is equated with objectivity. Piaget's notion of equilibration describes a mechanical self-regulation or 'psychological homeostasis' in which the organism not only accommodates itself to reality, but also assimilates portions of it into its own structure. Equilibration occurs, then, when something causes the subject's cognitive equilibrium to be undermined. For instance, a baby cannot assimilate all objects into its grasping reflex: an object may be too big, for example. This results in disequilibrium, and the child's action scheme – that is, the formal structure of the grasping activity pattern – must accommodate itself to this outer reality in order to restore equilibrium. The scheme thus becomes more complex. It is important that equilibration should result in a more stable equilibrium (it can then endure greater variation) which thus embodies more complex knowledge and is closer to objectivity.

These complementary processes are the beginning of the growth of cognition, which Piaget describes as follows:

All needs tend first of all to incorporate things and people into the subject's own activity, i.e. to 'assimilate' the internal world into the structures that have already been constructed, and secondly to readjust these structures as a function of subtle transformations, i.e. to 'accommodate' them to external objects. From this point of view all mental life, as indeed all organic life, tends progressively to assimilate the surrounding environment. This incorporation is effective thanks to the structure of psychic organs whose scope of action becomes more and more extended. Initially perception and elementary movement (prehension, etc.) are concerned with objects that are close and viewed statically; then later, memory and practical intelligence permit the representation of earlier states of the object as well as the anticipation of their future states resulting from as yet unrealised transformations. Still later, intuitive thought reinforces these two abilities. Logical intelligence in the guise of concrete operations and ultimately of abstract deduction terminates this evolution by making the subject master of events that are far distant in space and time. At each of these levels the mind fulfils the same function, which is to incorporate the universe to itself, but the nature of assimilation varies, i.e. the successive modes of incorporation evolve from those of perception and movement to those of the higher mental operations.

(1968b: 8)

This quotation not only describes Piaget's vision of the synthesis of genesis and structure, but it also locates the essential problem areas of the account: the separation of the self from external objects to form the concepts of the self and of the permanent object; the formation of primitive logical concepts such as those of class and order; and the formation of operational, that is, logically reversible, concepts such as class-inclusion relations between the class of counters and the subclasses of red counters and white counters. These three problem areas, which are, respectively, representative of the sensorimotor, pre-operational and operational stages, are united by the fundamental problem of how disequilibrium, which is essential for progress, is both recognized and responded to. As Piaget says above, the nature of equilibration changes through the various stages, although it fulfils the same function at each stage: the function is to restore equilibrium, but the way in which disturbance is dealt with changes; according to Piaget, cognitive development proceeds by the elaboration of schemes and

the continuous dual process of assimilation and accommodation which restricts the effects of disturbance and compensating reactions entirely to the levels considered; hence they become available for possible assimilation; what was ' disturbance on the lowest level becomes an internal variation of the system on the highest levels, and what was a

49

compensating reaction which resulted in cancellation finally plays the role of a systematic transforming agent for the variations in play.

(1978a: 70)

So, disturbance begins at the sensorimotor stage as physical obstruction (external to the subject), moves to perceptual disorder in the pre-operational stage (disturbance is partially internal), and finally appears as incompleteness of structures at the operational stage (totally internalized and anticipated variations). The equilibratory behaviours which compensate these disturbances (Piaget (1978a) labels them α, β, and γ behaviours respectively) are linked in an orderly and systematic progress. They do not appear as three distinct stages in Piaget's theory, although it may be said that each behaviour is characteristic of a particular developmental stage. Rather, α, β, and γ behaviours appear as phases which recur regularly throughout development, characterising three different degrees of equilibrium ranging from unstable equilibrium in a restricted field to a flexible and stable equilibrium.

So, in terms of disturbance and its recognition and transcendence, the three major problem areas for explanation are: (i) in the sensorimotor stage, the differentiation of subject and objects such that the disappearance of an object constitutes a disturbance; and the subsequent equilibration between action schemes and objects by means of compensation for physical obstruction; (ii) in the pre-operational stage, the formation of concepts by means of local trial and error correction for perceived disorder in, say, a collection of seriated items; and the subsequent internalization of action; and finally (iii) in the concrete-operational stage, the institution of operational reversibility by means of reflective abstraction of the structure of action in response to lack of coordination of internalized action schemes. This section concentrates on these three aspects of the equilibration model, and argues that Piaget does not succeed in producing a truly dynamic model of development. In all three problem areas he fails to describe the successive internalizations necessary for development in terms of the recognition and resolution of disturbances and the impetus behind equilibration to higher states; I will argue that his theory begs the question of how the individual alone can construct her world without foreknowledge of the structures in question. The problems illustrated here suggest that, as long as Piaget holds an assumption of the solitary knower, he cannot successfully describe the growth of understanding; I will pursue this idea further in Chapter 5.

Equilibration between action schemes and exterior objects:
sensorimotor intelligence and the concepts of
permanent objects and the self

For Piaget, an important feature of new-born infants is that they do not distinguish subject and objects: they have no consciousness of the self nor of a distinction between internal and external data; nor, he argues (1972a: 20) will they gain this understanding until there is an interest in others as persons (as opposed to the self), these being the first permanent objects. To account for the development of object permanence, Piaget argues, there is 'only one possible link – action – between what will later become differentiated into a subject and objects . . . the young infant relates everything to his body, as if it were the centre of the universe – but a centre that is unaware of itself' (1972a: 20–21).

Piaget notes the objection that it seems unlikely that a subject who is unaware of him/herself as the originator of actions should none the less centre those actions on the self. His counterargument is that these two features are compatible as long as actions remain uncoordinated:

each [action] constitutes a small isolable whole which directly relates the body itself to the object, as, for example, in sucking, looking, grasping, etc. From this there follows a lack of differentiation, for the subject only affirms himself at a later stage by freely coordinating his actions, and the object will only be constituted as it complies with or resists the coordinations of movements or of positions in a coherent system. On the other hand, so long as each action still forms a small isolable whole, their sole common and constant reference can only be the body itself, so that there is an automatic centring on it, although it is neither voluntary nor conscious.

(1972a: 21)

Piaget thus argues that there is an automatic centring on the physiological level. This seems reasonable, but it still cannot account for the development of a consciousness of self since, by Piaget's own account, consciousness presupposes reconstruction, and not a simple mirroring of physical existence. The subject's affirmation of the self through a coordination of actions presupposes a conscious self which Piaget has yet to account for.

Piaget argues that his claims here are supported by observation of the infant's behaviour during the sensorimotor period. There occurs, he says, 'a kind of Copernican revolution' (1972a: 21) as decentring of actions in relation to the body leads to the concept of the body as a separate object. As the child's self-awareness grows, these actions are coordinated; here there is some indication that, according to Piaget's account, coordination of actions does presuppose some degree of self-awareness since it involves a subject who

begins to be aware of himself as the source of actions and hence of knowledge, since the coordination of two of these actions presupposes an initiative which goes beyond the immediate behavioural interactions existing between an external object and the body itself.

(1972a: 22)

But, having said here that the coordination of actions presupposes some differentiation at least between subject and object, Piaget then goes on to contradict himself again by restating his original position that it is the coordination of actions that leads to the differentiation of self from objects:

> to coordinate actions is to displace objects, and insofar as these displacements are coordinated the 'group of displacements' which is thus progressively elaborated, makes possible the assignment of determinate successive positions to objects . . . . Such a differentiation of subject and objects entailing the progressive substantification of the latter, definitively explains the total inversion of perspective which leads the subject to consider his own body as one object among others, in a spatio-temporal and causal universe of which he becomes an integral part to the extent to which he learns to act effectively on it.
>
> (1972a: 22)

While it could be argued that Piaget is really describing an interactive process here which thus has no causal steps, he apparently intends that his description should have a clear direction as evidenced by the following: 'the coordination of the subject's actions . . . is the origin both of the differentiations between this subject and objects and of the decentring process on the level of physical acts' (1972a: 22).

Piaget (1978a) attempts some elaboration of this confusing account by setting the development of the concept of the permanent object in the context of the general development of the sensorimotor period. Recall that, at the beginning of development, objects are inseparable from the characteristics which link them to the body itself, for example, something is 'an object to suck'. This is a kind of 'reciprocal assimilation' of schemes, the earliest forms of which are the linking of vision and hearing (looking in the direction of a sound to find the corresponding visual image), and sucking and prehension (bringing to the mouth what is clutched and not part of the visual field). Later, there is coordination of vision and prehension (grasping what is seen). Piaget (1978a) argues that this comes about through the frequent experience of situations in which, for example, an object is both seen and heard, sucked or touched and seen, and so on.

These intersections of schemes (1978a: 90) produce a gap when one of them is activated without the other so that, for instance, when the subject hears an object but cannot see it, she is driven to look for it, so as to compensate the disturbance caused by the gap. In formal terms, the intersection of schemes presents objects with both x and y characteristics; in response to the disturbing contrary situation in which the objects are perceived as x without y, or y without x, compensation opposite to the disturbance is effected by a move to link x to y. This compensation constitutes what Piaget calls type α behaviour, and it entails a cancellation or neutralization of the disturbance.

The problem with this explanation is that it is difficult to see how it describes the creation of coordinations, since the disturbance felt by perceiving x without y must in fact be a product of the coordination xy (otherwise it would not be disturbing), and so it cannot be the only factor in the creation of xy, as Piaget seems to imply. Furthermore, as I have already observed, it is necessary to explain how this account can work in terms of disturbance without presupposing some object permanence (otherwise there is no necessary connection to be made between x with y or x without y). Thus the coordination described here depends on the notion of a permanent object, but the coordination itself is supposed to be effected without permanent objects, and is in fact part of the creation of permanence.

The question of the differentiation of subject and object is, apparently, never resolved. Leaving this problem aside, there remains a more general problem of the equilibration between action schemes and objects which centres on the question of how disequilibrium is recognized and acted on. Piaget (1978a, 1980) introduces the notion of the balance between affirmations and negations in order to clarify the nature of disequilibrium. Affirmation at the sensorimotor stage involves 'a complete taking over of the object's characteristics . . . without adding anything to them' (1980: 297). Negation consists in the recognition of what an object is not, and of the ways in which characteristics of the object cannot be assimilated into the action scheme in question.

When there is an imbalance towards affirmations at the expense of negations, contradictions (disequilibrium) occur: 'The reason for the contradiction stems from the fact that the subject, being centred on the goal or arrival point of actions because they are positive values, ignores the concomitant negations, subtractions or negative factors' (1980: 300). Thus disequilibrium is caused by the neglect of negative aspects of actions and objects. This is formalized by Piaget (1978a: 11) as follows: The subject has schemes A, B, C, etc. There exist exterior objects A', B', C', etc. Equilibration between the scheme A and objects A' demands that the subject's actions must not only possess characteristics a', but she must also distinguish these from different characteristics, non-a'. Similarly, in order to use the object A' with characteristics a', the subject must also use the scheme A, and not others which are non-A. Here, negative characteristics mark the limits of positive characteristics. They are shown to be necessary in the case in which a scheme A does not find its usual object A' to act on, and accommodates itself to object A'', which has the neighbouring characteristics a''. Successful accommodation gives rise to the modified scheme A_2; the initial state of A is subsequently denoted as A_1. Thus the scheme A now includes two subschemes A_1 and A_2 (so $A = A_1 + A_2$). For this subdivision to be balanced such that A_2 uses only object A'' and A_1 object A', the partial negations $A_2 = A$ plus non-A_1 and $A_1 = A$ plus non-A_2, are necessary.

A fundamental problem for Piaget is to explain the initial recognition of disturbance. In particular, he needs to show how a disturbance is noticeable in the sense that it produces an obvious obstacle. Ignorance, for instance, is not disturbing; one has to be thwarted, or recognize error in something one is doing or trying to find out, and so on, in order to recognize a problem or contradiction. Piaget's problem is, then, to show how the disequilibrium which results from an imbalance towards affirmations is recognized and resolved by the introduction into action schemes of the negative aspects of actions and objects. The problem of explaining recognition at the sensorimotor level is less serious than it is for the later stages: recognition is given by the immediate failure of physical action. However, recognition of failure does not automatically bring a realization of the reasons for that failure, and this is particularly true if Piaget's claim that development is both lawful and unpredictable is to be taken seriously: theoretically there is more than one outcome of disequilibrium, and so its resolution entails more than mere recognition of a problem's existence since, by Piaget's own account, it should entail reconstruction of the prior structures.

In this particular case, resolution entails recognition of the negative aspects of actions and objects. Piaget gives no account of how this comes about, implying instead that it is an automatic outcome of disequilibrium, and, moreover, that the new characteristics of objects are abstracted from the series of events in question. But this implicit abstraction model begs the question in that the abstraction of the relevant characteristics presupposes knowledge of those same characteristics; I will pursue this problem in the next chapter.

In summary, given Piaget's aim of accounting for knowledge without recourse to the idea of innate knowledge or the passive *tabula rasa* of empiricism, it is important that he explains the relationships between the coordination of actions, the conscious self, and the concept of object permanence so that, where it would appear that one must presuppose another, the genesis of the prior concept is well-explained. This Piaget fails to do. The central question to ask, however, is: As an account of development, is the assimilation–accommodation model a convincing alternative to empiricism and rationalism? Does Piaget really achieve a true 'middle way' by basing his account on activity? His intentions are clear: by basing knowledge on physical interaction with the world, Piaget hopes to avoid having to account for structure in an artificial manner, since he believes that this is inevitably given in the structure of action: it could not be otherwise, he thinks. Is this the case though? Piaget gives no adequate account of why action should be adequately characterized in terms of formal structure based on unconscious schemes 'biologically given'. Moreover, the emphasis on activity does not release him from having to say exactly what is assimilated to an action scheme: he still uses a notion of abstraction from objects which, as I will argue in the next chapter, presupposes knowledge of the object in question.

So, when he says (1972a: 23–4) that a child abstracts from objects the recognition of a suspended object as something to be rocked, Piaget still has to talk about the characteristics of objects, for these must somehow enter into the formalization of an action scheme. To be generalizable, an action scheme must contain some information about objects that fit the scheme, and this information would need to be of an abstract nature. By this account action schemes contain more information than Piaget would intend, and more than he can account for. This problem is the subject of the next chapter.

Equilibration between subsystems: concept formation and the internalization of action schemes in pre-operational intelligence

The elaboration of sensorimotor schemes follows the pattern of assimilation and accommodation towards greater equilibrium until the advent of language introduces a different quality into the child's intelligence such that actions can be represented in thought. While the child is at the sensorimotor stage, her intelligence is, as we have seen, immediate and rooted in action, these action schemes being thus unconscious (1972a: 25). Language, however,

> enables the subject to describe his actions. It allows him both to reconstitute the past (to evoke it in the absence of the objects which were previously acted upon) and to anticipate future, not yet executed, actions to the point where sometimes actions are replaced by words and never actually performed. This is the point of departure for thought.
>
> (1968b: 22)

The representation or 'interiorization' of actions cannot be totally straightforward though, and Piaget recognizes this. Conceptualization on the plane of thought is, then, of a different quality from the original action schemes that are being represented. Thus Piaget says that:

> It would be much too simple to hold that the interiorization of actions in the form of representations or thought, only involves retracing the course of these actions or imagining them by means of symbols or signs (mental images or language) without as such modifying or enriching them. In reality such interiorization is conceptualization, that is, it involves the transformation of schemes into concepts properly so-called, however rudimentary they may be . . . . Now, since the scheme is not an object of thought but rather the internal structure of actions, whereas concepts are manipulated in representation and language, it follows that the interiorization of actions presupposes their reconstruction on a

higher level, and consequently the elaboration of a series of novel features irreducible to the lower-level mediating structures.

(1972a: 26)

In his earlier work, Piaget is not at all explicit with regard to the process of reconstruction on a higher plane, and instead merely states the characteristics of pre-operational as opposed to sensorimotor intelligence. However, he attempts (1977, 1978a, 1978b, 1980) to elaborate on the nature of the internalization of action schemes by setting this within the context of the general process of equilibration of structures.

For Piaget (1972a: 28), a concept A is formed by the assimilation of all objects A into the same class with respect to their common property 'a'. This particular assimilation of objects to one another, which forms the basis of classification, also entails understanding of the quantifiers 'all' and 'some', an understanding which, together with a grasp of the nature of relations, pre-operational children lack. Thus, given some round red counters and some round and square blue counters, the pre-operational child declares that all the round counters are red, but denies that all the square ones are blue 'since there are also blue counters that are round' (1972a: 29). Similarly, she fails to recognize that, if $A < B < C$, the term B can be both bigger than and smaller than another term.

Piaget (1978a, 1980) describes the pre-operational child's progress towards concrete operational (reversible) concepts as due, once again, to the balance of affirmations and negations. Affirmation relates to the positive characteristics of classes and relations in terms of the common characteristics of their members, or the links between them, while negation consists in disallowing an object's inclusion in a class or participation in a relation. A pre-operational imbalance towards affirmation leads to contradiction between subsystems and a lack of coordination which will remain unresolved until the operational stage.

Thus the coordination in question entails recognition that any necessary construction has a negative aspect – to assert the necessity of $A < C$ is to exclude $A \geqslant C$. Piaget (1980: 303) declares that 'it is neglect of the negative aspects . . . that, by compromising the compensations indispensable to the coherence of the whole, account for . . . contradictions involving relations between sub-systems or between schema'. Thus, if there are two subsystems, S_1 and S_2, Piaget (1978a: 11) argues, an intersection of these will require negations such that what is common to S_1 and S_2 – that is $S_1 \cdot S_2$ – is opposed to $S_1 \cdot \text{non-}S_2$ and to $S_2 \cdot \text{non-}S_1$: the coordination requires partial negations.

In comparison to the first form of equilibration, the negation necessary for restoring balance at this stage is no longer simply practical but constative in nature, and its role becomes less negligible as the conceptual framing of objects permits the integration into these interpretative systems of an increasing number of external perturbations, which then take the form of functional variations which must be considered in their own right and no longer simply eliminated.

(1980: 298)

This is type β behaviour, which consists in integrating the disturbing element into the system so that

> the unexpected fact is made assimilable . . . the classification will be recast to coordinate the new category with the others; the seriation will be extended or distributed in two dimensions, etc. . . . What was disturbing becomes a variation within a reorganised structure, thanks to the new relations which make incorporating the element possible . . . . By integrating or internalising the disturbances at play in the cognitive system, these type β behaviours transform them into internal variations.

(1978a: 67–8)

The equilibrium thus achieved is not wholly stable; Piaget (1980: 298) makes a distinction between this local conceptual framework, and general operatory structures: the number and quality of pre-operational negations remain inferior to those of affirmations such that negation is still transitory, lacking the durability of negation in reversible operations.

While the importance of negation in Piaget's system is clear, the origin of negations and the exact nature of disturbances are not as yet: the notion of noticing what something is not, or what an action is not, is problematic. However, Piaget attempts an explanation in these terms by means of two mechanisms: the repression of unconscious perceptions and the repression of know-how.

The first of these two mechanisms concerns the child's early conceptualizations which, Piaget maintains, are frequently distorted because they are either erroneous or incomplete. If a concept is simply erroneous because, say, one of the quantifiers is wrongly applied, there is a possibility that the error may come to light at some point and become apparent to the child. But if the distortion is due to an omission, its resolution is not so easily explained, since the omission may not be apparent to the subject, and thus need never be corrected: a child who attributes constant width to a clay sausage during its extension can continue to neglect its thinning. This presents a major problem for Piaget because omissions must be noticeable in order to trigger equilibration: 'at first glance the regulations they create seem only to fill gaps, and in terms of equilibrium a gap constitutes a disturbance

57

involving a compensation reaction only insofar as it corresponds to an already active scheme' (1978a: 143).

Piaget must, then, explain these gaps, and he does so by arguing that they are not so much gaps as repressions, by which certain elements are set aside by the assimilating scheme. This is explained by the notion that the perceivable characteristics may be unconsciously perceived (subception), or they may be momentarily conscious and subsequently lost due to lack of integration into the conceptualized awareness. But, Piaget argues, the perceptible content, whether it is perceived or not, is always present, having a potential influence on the constraints exercised by the assimilating schemes. The equilibrium achieved between the conceptualized form and the perceived content is thus unstable, and Piaget's normal equilibration model follows on:

> since it is no longer a question simply of filling gaps but of lessening the repression of the elements which earlier were set aside, the disturbance can be attributed to the nascent power of these elements, which tend to penetrate into the field of recognised observables, and the compensation will then consist in modifying the disturbance until it becomes acceptable.
>
> (1978a: 146)

The second mechanism concerns the subject's awareness of actions. Here the content is the set of sensorimotor processes which produce action, and the form is the system of concepts used by the subject to become aware of these. Piaget (1977) observes that children can achieve in action quite complex coordinations (for instance, releasing a stud from a sling so that it reaches a target) without being aware of what they do, or even misconceptualizing this. There is, therefore, a mismatch between know-how and knowledge-that.

Piaget argues that, when the subject misconceptualizes what she is doing, she is not simply neglecting elements or making incomplete conceptualizations: the gaps in her awareness indicate a repression of certain perceptions because they are contradictory to a frequently used conceptual scheme. How, then, is the subject, successful as she is in the task, to arrive at a true conceptualization of such actions? Piaget suggests that in this situation, an unstable equilibrium is achieved in which certain elements of the content are rejected by the form, but the content exerts pressure in opposition to resistance from the form. Thus the situation is parallel to that for objects:

> certain schemes, used in a sensorimotor manner but not conceptualised, exert pressure on the conceptualisations and tend to shake their repression. Therefore, the object, which is manipulated in a sensorimotor manner before lending itself to the conceptualisation, includes a certain

number of latent or virtual observables due to (or made sensitive by) this sensorimotor manipulation (therefore the likelihood of the existence of 'subceptions') which exert pressure against the repression.

(1978a: 151)

In conclusion, Piaget argues, awareness of these already successful actions proceeds from the periphery of the action (where contacts are made with the object) to the centre (the internal mechanisms of the action).

Thus Piaget relies on a general model whereby unconscious perceptions and know-how are repressed in order to solve the problem of the recognition and transcendence of disequilibrium. Where there was, apparently, a gap or omission in knowledge, which would create a considerable problem of explanation for Piaget's theory, Piaget instead posits a conflict between 'repressed' knowledge and the subject's conceptual scheme. But this begs the question; Piaget's use of the idea of repression involves an implicit identification of know-how with knowledge-that, which is necessary if know-how is to be of a form to act on the subject's present (mis)conceptualization. But, for know-how to become knowledge-that, the subject must become cognizant of know-how, which, by Piaget's own account, must therefore be reworked: cognizance involves reconceptualization. Piaget's account begs the question because it is this very cognizance that is both the source and the result of awareness of disturbance.

There is a parallel problem on the level of objects. Piaget wants to argue that, in the same way that know-how is repressed, so unconscious perceptions are repressed too. Again, there is the problem of how such perceptions can have the force to undermine an established concept without begging the question of how recognition of the perceptual disturbance is possible without presupposition of the concept in question. It seems that Piaget cannot account for the introduction of classes of objects related by a certain common property without begging the question of how the subject comes to use those classes in the first place. Piaget himself remarks on the problem of consciousness with regard to pre-operational contradictions:

these contradictions stem from a lack of coordination, and for that reason remain unconscious for a fairly long time, since transcendence of them cannot be achieved other than by the intervention of overall operatory structures, whose common characteristic is their intrinsic necessity.

(1980: 302)

And elsewhere he admits that 'consciousness of a contradiction between schemas is not produced until the level at which the subject becomes capable of transcending it' (1980: 292). I will discuss these problems further in the following section.

Equilibration between subsystems and totalities: the completion of concepts by reflective abstraction in operational intelligence

As the description of pre-operational thought suggests, the major feature of the concrete operational period is that the interiorized actions of the pre-operational stage become operational, and are thus completed. Prior to operations, there is a period of imbalance towards affirmations which leads to contradictions resulting from erroneous inferences and false implications. Thus if there is a class of red cubes A all containing a bell B, imbalance towards the affirmation $A \Rightarrow B$ (A implies B) causes the subject to infer wrongly that $B \Rightarrow A$ (all cubes containing a bell are red), forgetting that if $A \Rightarrow B$, then it is possible to have $\overline{A} \cdot B$ and $\overline{A} \cdot \overline{B}$, but not $A \cdot \overline{B}$, which is the negation of $A \Rightarrow B$. Piaget maintains further that there is a partial negation such that, if a child is sorting a set of round counters B into whites A, and reds A', she will say that the latter are 'not white', but she has not necessarily constructed the secondary class of 'non-white rounds' (A' = $B \cdot \overline{A}$), and thus is unable to perform the related class-inclusion task. Piaget argues that:

> we have to wait until the operatory level before we find negation being correctly manipulated in such a case; and this remains true of all the other 'groupings' of concrete operations . . . . The operatory reversibility . . . consists of bringing a reverse operation, and thus a negation, into correspondence with each direct operation or affirmation.
>
> (1980: 297)

Affirmations at this stage involve class-inclusion relations, for instance, while negation, on the other hand, entails the recognition that for every class A there is a complementary class non-A. Achievement of the necessary balance between negation and affirmation and transcendence of contradictions entails that reverse operations 'take in and make over, in the form of internal variations within the system, that which until now had remained partly in the form of external perturbations' (1980: 299). This behaviour Piaget terms 'γ behaviour' (1978a: 68-9), and it has an anticipatory quality such that variations are foreseen and deduced and so lose their disturbance characteristic: what would have been disturbances enter the system as transformations capable of being inferred. Here again is the closure of systems described in Chapter 3, which 'eliminates any contradiction emanating from without or within' (1978a: 69).

Where does γ behaviour get its anticipatory quality from, and what is the impetus behind the equilibration which takes the subject's thought to this level? Piaget maintains that, just as the repression of particular elements exerts pressure on conceptual organization in the pre-operational stage, so in this later stage 'quantities of approximate knowledge and of poorly solved problems . . . constantly inspire the research. It is therefore in this region

60

that the unstable play of dynamic processes develops and makes valuable certain questions while brushing aside others though not cancelling them' (1978a: 176).

So Piaget finally places the onus for search on the conscious individual; as Hamlyn (1978: 54ff.) comments, for Piaget, 'the individual constructs his world'. But Piaget's 'commonsense remarks' (1978a: 176) cannot account in any detail for how the child develops the anticipatory behaviour which carries her from the level of trial and error to operational seriation or classification, or how she achieves awareness of her own successful actions. As Piaget says, the subject is not troubled by her ignorance, and as long as it is up to the subject alone to reflect on her own action, it is hard to see what will bring this about. Like the conceptualization of the pre-operational stage, the impetus for this reflection relies on a 'disturbance':

> We never abolish a question; even thrust back, it arises by itself because of its implications which bind it to what appears resolved. It is therefore by no means tautological to imply that each new construction, including inferential coordinations, which arise from some area of study . . . aims to compensate not deficits or random gaps but those which are part of already activated schemes, hence are disturbances which until now prevented the solution of particular problems.
>
> (1978a: 176)

Conservation, for example, is

> the result of equilibrium, in the sense in which the subject's reactions compensate for external changes (for example by imagining transformations in the inverse direction) . . . . [The conservation of length] only occurs at a stage where he begins to reason in terms of reversibility . . . . As for knowing how this latter self-evidence is acquired . . . the general answer will again be based on a reference to a structured whole, which achieves the equilibrium progressively as a result of the demands of internal consistency and of compensations, finally arriving at meanings satisfying to the mind and at the same time compatible with experience.
>
> (1966: 196)

But Piaget is vague as to the details; ultimately, he only describes the end products of equilibration, and not the process itself. He does not explain, for instance, how the 'disturbance' which prevents solution of a class-inclusion problem is in fact connected to the lower-level classificatory schemes, or how it is eventually recognized and compensated without presupposing the structure which makes the compensation. The question of circularity in Piaget's theory, raised by Piaget himself as remarked on page 59, remains.

Problems of disturbance aside, the question of circularity still persists in Piaget's account of reflective abstraction. The theory suffers, as signalled in Chapter 2, from the problem of how to describe different qualities of thought such as are represented by the transition from pre-operational to operational thought. Recall that it is Piaget's anti-empiricist contention that active inter-pretation accompanies all experience, and that the results of physical experience must be apprehended through the medium of a logico-mathematical framework, even if it is of the most elementary kind. The case of logico-mathematical experience is rather more complicated though; Piaget notes that there is an apparent circularity: logico-mathematical experience involves noting both the physical properties of objects and the results of action (in terms of new properties which result from actions such as 'being ordered'). But, if 'noting' these (physical) properties already presupposes logico-mathematical frameworks, the problem then arises: does noting the results of action presuppose the very framework which is to be abstracted? Piaget avoids the problem by giving a qualified answer which is that the two frameworks – the preliminary and the abstracted – are of a different quality, from successive planes of thought:

To derive new knowledge from one's own actions consists not merely of becoming conscious of a preliminary organisation without modifying it other than through the passage from unconsciousness to consciousness, but rather of generalising this preliminary organisation and of representing it, in the psychologist's sense of the term, in the form of a larger model of operations capable of being conceived simultaneously. A scheme of action is, in fact, only the form of a series of actions which take place successively without a simultaneous percep-tion of the whole. Reflective abstraction, on the other hand, upgrades it to the form of an operational scheme, that is, of a structure such that when one of the operations is used, its combination with others becomes deductively possible through a reflection going beyond the momentary action.

(1966: 237)

So, Piaget states the process of deriving new knowledge from the results of actions or coordinations of actions as follows:

(a) logico-mathematical experience consists of observing the results of action performed upon any object; (b) the results are determined by the schemes of the actions thus carried out on the objects; (c) but in order to observe (or to 'note') these results, the subject has to carry out other actions (of 'noting') using the same schemes as those the product of which must be examined. However, (d) the knowledge acquired is new for the subject, that is (although in principle, a simple deduction might

62

have replaced experience) experience teaches him what he was not aware of in advance. We must therefore conclude (e) that the abstraction by means of which the subject acquires new knowledge of the results of his actions – knowledge which is new for his consciousness – involves some construction; and this has the effect of translating the scheme and its implications into terms of pre-operations or conscious operations, the later handling of which will allow him to replace by deductions the experiences or empirical procedures which have thus become useless.

(1966: 237)

In other words, reflective abstraction is an account of how the form of one stage becomes the content of the next; for instance a child becomes conscious of the existence of order among objects due to the 'reconstruction on a higher plane' of her previously purely empirical ordering actions. As to how this happens, however, Piaget is extremely vague here. He avoids the issue of the actual mechanisms of reflective abstraction, and gives a purely formal account of stage A and its successor stage B, falling back on the question-begging 'we must conclude . . . that the abstraction . . . involves some construction'. Piaget 'must conclude' this only because he has already taken a stance against empiricism and abstractionism, not because his description of what is happening leads logically to this. Piaget's account is still circular; in order to abstract knowledge from her actions, the child has to have some structure already in order to be in a position to perceive the patterns in her actions (otherwise she would be behaving as the empiricist says she does), but since Piaget wants to argue that the knowledge that the child gains at this point is genuinely new, he has to say that the abstraction is of a special kind which does in fact entail that the new knowledge is really new in a qualitative sense. But this is so only because Piaget says it must be, and not because evidence or logic demands it. He does not avoid the problem of how to explain exactly how the reconstruction is achieved.

Closely connected with the problem of circularity is that of goals and Piaget's ultimate attribution to the subject of behaviour which seeks to close the gaps in her own structures. In asserting that the individual experiences such gaps and actively tries to close them, Piaget implies an unacceptable degree of self-consciousness which does not fit with the notion that development is the outcome of unconscious reflective abstraction onto a higher plane. The elaboration of forms cannot be associated with a 'need' for differentiation and integration since, as Piaget himself comments, this would only displace the problem to the causes of the need and beg the question as to why the two processes need to be balanced. Piaget's alternative solution involves the notion, first discussed in Chapter 2 (pp. 24–5), that no structure is complete or understood until integrated into the next, such that progress comes about through a property that is intrinsic to development itself:

We believe that we have verified the constant unity of constructions and compensations . . . . New possibilities are revealed by the constitution of a structure, which shows virtual disturbances as compared to its real state, but which can be compensated by coherent incorporations . . . . Our proposition is therefore that operations dealing *with* preceding operations owe their creation to these situations, and that the extension of the previous system consists of an improved equilibration insofar as the disturbance arising from a virtual modification is surmounted by such an incorporation.

(1978a: 180–1)

Thus a central idea of Piaget's account is that cognitive development contains its own drive forward; that one construction leads to the anticipation of another by means of a kind of inertia of incompletion. Thus Piaget (1978a: 182) refers to conscious feelings that 'something remains to be done . . . we haven't exhausted the possible ways of reaching a goal opened by a structure completed elsewhere'. But such an inertia, stated as it is on this intuitive level, does not fit with the formally stated theory of assimilations and accommodations, and it is unsatisfactory in this respect. There is a mismatch between the structural evens and the subject's awareness: on the one hand, Piaget's theory rests on the notion that no structure is truly complete until reflected onto a higher plane and thus integrated into the next, while on the other hand, it posits a conscious sensation of disturbance, and even incompleteness. But the link between the structural events and the subjective awareness cannot be one of simple mirroring. Awareness, by Piaget's own account, is the product of reconstruction, which means that, as Piaget himself observes, transcendence of one stage is constituted by, and constitutes, arrival at the next.

On a more general level, there is a question to be asked of any theory of development which involves abstraction, even reflective abstraction: For what reason should an individual abstract either the underlying form of abstractions or the properties of objects? It can be argued, and this is elaborated in the next chapter, that such abstraction presupposes the knowledge in question. This is an argument not unlike that used by Piaget himself, in the passages above. However, Piaget clearly believes that by stipulating a different quality of thought for that which is presupposed and that which results he avoids the problem. Whether he does in fact do so is again a problem I will consider in the next chapter.

It is clearly vital to Piaget's theory that action has structure and that any action is adequately described in these terms: just one formal structure need be abstracted from any one action. It is by virtue of this assertion that Piaget escapes from many of the criticisms that could be levelled against his theory. If there is only one necessary and sufficient formal structure for any one action, then Piaget cannot be accused of assuming innate structures in his theory: the important role played by action in the theory and the assumption of the universal form of action releases him from the necessity of explaining why that particular structure arises, and leaves him only with the problem of explaining how it is discovered, or rather how it is brought to consciousness by means of experience. Not all knowledge is amenable to this analysis, however. In the following chapters I will argue that, just as ordinary abstraction presents a problem in its assumption that a concept – say, 'redness' – can be grasped by mere repetition of the experience of things which are red, so it is possible to argue that there is no necessary reason why repeated experiences of actions, even if these do have an essential formal structure, should lead to learning concepts which have the same formal structure.

Piaget's account of the individual's growth towards mathematical knowledge is one which relies on distinct mechanisms of change: assimilation, accommodation, and equilibration, which are quasi-biological notions, and the coordination of actions and reflective abstraction, which again have neurological roots but are driven by equilibration. The criticisms of these mechanisms made in this chapter – that Piaget does not explain how an event is perceived as disequilibrating or how equilibration results in higher forms of equilibrium, or what the significance is, if any, of the essential formal structures of action and knowledge – are developed in the next chapter, beginning with the notion of the underlying structure of knowledge and the problem of growth from weaker to stronger logics.

5
Does Piaget give an adequate account of growth?

In Chapter 4 I discussed Piaget's account of the growth of understanding, and introduced criticisms of the mechanisms of growth that he proposes. The most important criticism is that Piaget's account entails that the individual alone must recognize and act on contradictions, and that it begs the question of how he can transcend those contradictions without foreknowledge of their resolution on a higher plane. This chapter enlarges on this criticism and suggests that an account such as Piaget's entails inherent and unavoidable problems in explaining the growth of understanding. I will argue that these problems have their root in the fact that Piaget assumes an essentialist account of knowing; that is, he characterizes knowing about number in terms of certain necessary and sufficient conditions, specifically the grasp of the logical notions of class and order. Relatedly, Piaget assumes that all the uses of number have a common factor such that understanding of the number concept in one situation entails understanding it in all situations. I shall return to a more detailed discussion and criticism of Piaget's essentialism in Chapter 6.

A common argument against Piaget is that he wrongly excludes social factors from his account of development; in fact, Piaget foresees the argument (see Piaget 1966, for instance), and argues coherently against it. Given his formal characterization of knowing number and his emphasis on the structure of action as the basis of knowledge, Piaget's rejection of environment or society as dominant factors in the growth of knowledge in the individual is sound. However, I will argue in this and subsequent chapters that a reconceptualization is possible which extends the role of social factors into the characterization of knowledge itself, not just its growth, and so allows for an account of development which is more successful than Piaget's. Piaget's assumption of a solitary knower who must construct his world can be replaced with an account of an essentially social being for whom knowing number involves entering into the social practices of its use. This chapter presents the first part of this argument. The first section examines Fodor's (1976) criticism of Piaget's account in terms of the plausibility of a transition from weaker to stronger logics. It is suggested that Fodor's claim that cognitive development must necessarily be a question of the maturation of innate ideas is not the only solution to the problem of explaining development for Fodor assumes, like Piaget, an essentialist characterization of knowledge. The second section considers Piaget's reflective abstraction model in terms of whether it can avoid the problems of an ordinary abstractionist account; I will suggest that, since not all actions can be adequately described in terms of structure alone, Piaget's account fails in this aim. The idea that an action which can be described as, say, putting objects into one–one correspondence, is only understood as that particular action when it occurs in a particular social context leads to the claim in the third section that, as Hamlyn (1978) argues, a successful account of the growth of understanding entails the view that knowledge is intrinsically social.

The transition from weaker to stronger logics

Piaget's theory depends for its 'middle way' explanation on the notion of growth as a transition from weaker to stronger logics. It is in this fashion that he blocks the infinite regress inherent in rationalist accounts of knowledge such as Chomsky's. For the rationalist, accounting for the origin of knowledge is a major problem; according to Piaget, Chomsky's account of language learning is forced to postulate innate knowledge because it does not describe a process of qualitative change in knowledge during the development of the individual. Whereas Piaget attempts to show how knowledge develops by means of reconstruction and qualitative change, the Chomskyan growth process is one of maturation of pre-existing full-blown innate principles. Piaget's account entails that later knowledge emerges from earlier knowledge and is therefore not identical with it. He thus avoids the need to propose complex innate knowledge in the child, but is left instead with the problem of how to explain how later stages emerge from earlier ones with which they have links, but no identity. This section concerns Fodor's (1976) argument that it is not possible to generate stronger from weaker logics, which he intends as a direct criticism of Piaget.

Fodor starts off with the basic premise that it is impossible for someone to learn a language which is logically more powerful than the one he already knows (1976: 86). It is, however, possible to use part of a language to learn another part of the same language, so that, in using a dictionary, for instance, one can understand the word W by means of the definition D if the following conditions are met (1976: 87):

(i) 'W means D' is true
(ii) one understands D

Now, Fodor argues, if (i) is satisfied, then D and W must at least be coextensive (that is, the two terms have the same extent of application, or they refer
to the same objects), which means that if (ii) is true

> someone who learns W by learning that it means D must already
> understand at least one formula co-extensive with W, viz., the one that
> D is couched in. In short, learning a word can be learning what a
> dictionary definition says about it *only for someone who understands the
> definition.* So appeals to dictionaries do not, after all, show that you
> can use your mastery of a part of a natural language to learn expres-
> sions you could not otherwise have mastered. All they show is what we
> already know: once one is able to express an extension, one is in a
> position to learn that W expresses that extension.
>
> (1976: 87)

Transposed into the field of cognitive development, this argument has
considerable consequences for Piaget's claim that development is a question
of changes in competence rather than performance, that is, that the child's
thought progresses from weaker to stronger logics. Although Fodor accepts
the possibility of a series of stages in development which differ in terms of
the expressive power of their underlying logics, he argues (1976: 89) that
concept learning cannot provide the mechanism of transition from stage to
stage. He reasons that:

(i) If a stage one child is trying to learn the concept C, at the least
she has to learn the conditions under which something is an instance
of C. So, if F is coextensive with C, the child has to learn that there
is an x such that x is C if and only if x is F.
(ii) To learn this (that is, (x) (x is C if x is F)), the understanding
of F is presupposed.
(iii) However, if C is a stage two concept, then it cannot be coextensive with any stage one concept. Otherwise, there would not be a
difference between the two stages in terms of the expressive power of
their underlying logics.
(iv) Therefore, the stage one child cannot represent the extension of
C since there is no coextensive concept available to him within the
stage one system.
(v) Therefore the stage one child cannot learn C.

So, for instance, in Piaget's system, the pre-operational child cannot grasp
the concept of conservation because understanding of this concept presupposes operational thinking which, by definition, is not available to the pre-
operational child. However, Fodor argues,

if the child cannot so much as *represent* the conditions under which quantities are conserved, how in the world could he conceivably learn that those *are* the conditions under which quantities are conserved? Small wonder that Piaget gives so little by way of a detailed analysis of the processes of 'equilibration' which are supposed to effect stage-to-stage transitions.

(1976: 90)

A second example is Piaget's (1954) description of the child's construction of space, a concept which becomes more complex over successive stages; in this case, stages are distinguished from one another not in terms of what concepts can be expressed but in terms of the range of application of the same concept. Fodor claims that the same argument applies, however, since

learning that the concept C applies in the domain D is learning that there are individuals in D which do (or might) fall under C. But, by assumption, learning *that* is a matter of projecting and confirming a hypothesis, viz. the hypothesis that $(\exists x)$ (x is in D and (possibly or actually (Cx))). Trivially, however, one cannot project or confirm that hypothesis unless one is able to represent the state of affairs in which some individual in D satisfies C. So, again, learning does not increase the *expressive* power of one's system of concepts (construed as the set of states of affairs that one can represent).

(1976: 92–3)

As a final illustration of his point, Fodor takes as exemplars two real logical systems – standard propositional logic and first-crder quantification logic – and sets a hypothetical computer programmed with the first system the task of learning the second. The two systems differ in that first-order logic is stronger than propositional logic because every theorem of the first is a theorem of the second but not vice versa. Fodor argues that his computer could not learn first-order logic because the comparative weakness of propositional logic would prevent it from representing certain formulas and conditions which belong to first-order logic.

So, he claims, neither a person nor a machine can use an existing conceptual system in order to learn a more powerful one: 'What couldn't happen is that it gets from stage one to stage two by anything that we would recognize as a computational procedure. In short, trauma might do it; so might maturation. Learning won't' (1976: 93). Fodor's argument amounts, then, to saying that if there are stages which differ according to the expressive power of their underlying logics, then there can be no such thing as 'learning a new concept': if learning C involves the acceptance of a particular hypothesis, then the concept C is presupposed in projecting that hypothesis.[1] The only coherent solution, he argues, is one which 'presupposes a very extreme nativism. And this may not be so bad as it seems' (1976: 96).

Fodor's general argument against Piaget is that he gives no concrete explanation as to how stronger logics can be generated from weaker. Knowledge, characterized as it is by Piaget in terms of logical prerequisites and necessary and sufficient conditions cannot, in point of logic, give rise to genuinely new knowledge previously inexpressible in the child's prior logic/language. Piaget, of course, is aware of the difference that Fodor points out between logics of increasing strengths, and he is concerned to emphasize that the stronger logic is indeed irreducible to the weaker. This is an integral part of his epistemology, as Chapter 2 showed. He does not, however, advocate a collapse back into nativism as Fodor does, but cites reflective abstraction as the special means by which qualitatively new knowledge can emerge from old. The next section considers this claim; I will argue that the notion of reflective abstraction cannot fulfil Piaget's purpose, and that it is necessary to characterize what is to be learnt in social terms, since it may be that this is the only means by which the problem of how knowledge grows can be solved.

Reflective abstraction and the growth of knowledge

This section continues the criticisms of reflective abstraction which I began in Chapters 2 and 4, concentrating here on the validity of Piaget's general idea of the generation of qualitatively new knowledge from prior structures. I will argue that Piaget's important assumption that all actions can be described in terms of single structures is questionable, and that the mechanism of reflective abstraction fails to make any significant departure from other accounts of growth, and in fact presents no real alternative to ordinary abstraction.

It is important to distinguish between Piaget's notion of reflective abstraction and the simple or Aristotelean abstraction of empiricism. Thus Piaget writes:

[Reflective abstraction is] characteristic of logico-mathematical thought and differs from simple or Aristotelean abstraction. In the latter, given some external object, such as a crystal and its shape, substance, and colour, the subject simply separates the different qualities and retains one of them – the shape, maybe – rejecting the rest. In the case of logico-mathematical abstraction, on the other hand, what is given is an agglomeration of actions or operations previously made by the subject himself, with their results. In this case, abstraction consists first of taking cognisance of the existence of one of these actions or operations, that is to say, noting its possible interest, having neglected it so far . . . . Second, the action noted has to be 'reflected' (in the physical sense of the term) by being projected onto another plane – for example, the plane of thought as opposed to that of practical action, or the plane of abstract systematisation as opposed to that of concrete thought (say, algebra versus arithmetic). Third, it has to be integrated into a new structure . . . which must first of all be a reconstruction of the preceding one, if it is not to lack coherence . . . it must also, however, widen the scope of the preceding one, making it general by combining it with elements proper to the new plane of thought; otherwise there will be nothing new about it.

(1971a: 320)

Reflective abstraction is special because what is abstracted is not qualities of objects, but the structure of actions; development onto a higher plane of thought involves conscious recognition of the structure in question. However, I will argue in this section that Piaget's notion of reflective abstraction is not substantially different from ordinary abstraction and thus is subject to the same criticisms. The first part of this section lists these criticisms: simple abstraction fails to account for learning because it demands implausible psychological abilities, and because it assumes that having a concept entails the recognition of essential recurrent features in experience. Apart from the fact that, as I argued in the previous section, such a system can only work if the concepts to be abstracted are presupposed, an abstractionist theory has particular difficulty in accounting for the learning of logical concepts such as 'some' and 'not' and relational concepts such as 'big/small' and 'right/left'. The second part of this section compares reflective abstraction and simple abstraction. In the same sense that abstractionist accounts assume that concepts consist of essential recurrent features such that there is an essence of 'redness', for instance, Piaget assumes that actions are sufficiently described in terms of structures. Simple abstraction relies on abstraction of these essential features, while reflective abstraction assumes the abstraction of structure. Piaget's assumptions about the structure of action are questionable, and I will argue here that his account is vulnerable to the same criticisms as simple abstraction is.

Simple or Aristotelean abstraction as an explanation of growth

Commonly, abstraction is taken to be the formation of concepts by the abstraction of features directly given in sense experience – Piaget's 'Aristotelean abstraction'. This is the simplest form of abstraction, and it will be useful to consider this before turning to the abstraction of structure from actions. I will draw here on Geach (1957), who presents a standard argument against abstractionist accounts. Simple abstraction as an account of learning is normally applied to concepts like 'red' and 'round', which refer to simple

71

sensible qualities. The process of learning itself is presumed to be one of ostensive definition, that is, one learns a concept by means of direct and repeated exposure to the object; for example, one learns the concept 'red' through the experience of the recurrent labelling of red objects as 'red' by means of pointing and the like. This argument is problematic, however; there is no necessary reason why a person should single out the particular quality of redness rather than any other, and, arguably, doing so would in fact presuppose possession of the very same concept.

There is an alternative abstractionist argument put forward by Price (1953) which supposes that we might be more likely to learn words such as 'cat' by a double process of abstraction where several utterances which are otherwise unalike have in common one feature – the word 'cat' – and these are connected with several environmental circumstances which have in common one factor – the presence of a cat – but are otherwise different from each other. The most striking feature of such a concept-learning theory is its sheer implausibility in terms of the psychological demands it makes on memory alone. Additionally, the theory does not explain how, in the first place, a person can pick out the relevant features of any situation or piece of speech, since both are highly complex. This argument applies even in the simplest of circumstances – that is to say, when someone is talking about the present situation – but it has even more force when applied to the less simple description of things not present, past events, and so on. As Geach notes:

> we can use the terms 'black' and 'cat' in situations not including any black objects or any cat. How could this part of the use be got by abstraction? And such use is part of the very beginnings of language; a child calls out 'pot' in an 'environmental situation' in which the pot is conspicuous by its absence.
>
> (1957: 34–5)

The idea of simple abstraction not only presents a problem with regard to how concepts are learned; it also assumes that concept learning entails the recognition of recurrent features. However, it may be the case that there are no recurrent features of objects such that all and only objects with that feature fall under a concept, and a successful argument against this assumption would make a strong case against an abstractionist model of learning.

The abstractionist model of learning has special problems over and above these with respect to logical and relational concepts. Logical concepts such as 'some', 'or', and 'not' do not of course refer to sense experience – there are not examples in the world of alternativeness or negativeness from which to abstract the concepts 'or' and 'not'. A plausible abstractionist argument, then, is to describe logical concepts not as separate concepts corresponding to the logical words, but as special uses of non-logical concepts. So, for instance, it can be said that in saying 'not red' a person is simply using the concept 'red' in a special way; they are making reference to one concept only: 'red'. Thus an abstractionist argument of this sort is only bound to explain concepts expressed in positive terms. There is, however, a counterargument suggested by Geach (1957: 25–6) which is that the logical concept 'not' is in fact used as a distinct concept as it appears in the phrase 'if every P is M, and some S is not M, then some S is not P': since 'M' does not express a concept, the 'not' in 'not M' cannot be a mere reference to a special use of the concept 'M'. 'Not', then, must refer to a distinct concept and is not, therefore, explained by the abstractionist model.

There is a different problem in the case of relational concepts; taking the relation right–left, it is the case that the relation 'to the right of' cannot exist or have meaning without its converse 'to the left of'. This poses a problem because if an abstractionist account looks for the recurrent features of pairs of objects that stand in this relation to each other, the most that can be abstracted is the notion of right–left ordering. Hence if abstraction relies on looking for one recurring feature of a number of situations/objects/utterances which are otherwise different, then this will not lead to the ability to tell which thing is to the left and which is to the right, because there is a right object and a left object in all instances. The two halves of the concept act as though they were two features, so that one recurring feature alone can never be distinguished. So the most that an abstractionist theory can account for is the concept of right–left ordering. Even this is unsatisfactory, however, since the concept of 'right and left' itself presupposes an ability to tell which thing is to the left and which is to the right (Geach 1957: 33).

There are a number of problems, then, with the notion of learning by simple abstraction, not least the assumption that there exist recurring features of objects such that all and only objects with that feature fall under a particular concept. Geach's general point against abstraction is, furthermore, that abstraction can only work if the very concepts to be abstracted are presupposed, while the case of logical concepts presents a particular problem in that the abstractionist model presents no account at all of how they are learned. In the following part I will examine whether or not Piaget's particular version of abstraction – reflective abstraction – succeeds in avoiding these problems.

Reflective abstraction as an explanation of growth

Piaget's major innovation in proposing a theory of reflective abstraction is his claim that the material for abstraction lies not in objects but in the subject's actions on objects. As I showed in Chapter 4, its major application is the development of increasingly powerful logical systems including, of special interest here, the reversible logic of classes and relations. So, to recap briefly, reflective abstraction involves a series of steps such as the following:

(i) the subject uses action schemes which are unconscious, and does not notice their implications;

(ii) he notices (and in so noticing uses the very scheme to be abstracted) the implications of his actions;

(iii) he abstracts the necessary logic of the scheme in question and reconstructs it;

(iv) the subject now has new knowledge of his actions.

So, for instance, the child passing from the stage of pre-operational thought to that of concrete operations derives basic seriation from partial orderings occurring in the construction of empirical pairs, triplets, and series. In constructing these empirical pairs and so on, the child has used order relations in placing A to the left of B, but does not know this consciously, and correspondingly does not know that the relationship can be seen in terms such that B is to the right of A. Similarly, in constructing empirically the triplet $A<B<C$, he does not know that B can be at the same time bigger than and smaller than another term in the series. However, at some point the child notices the implications of these ordering actions, and abstracts the logic that if $A<B$, then $B>A$ and if $A<B<C$ then $B>A$ and $B<C$.

There are two major criticisms to be made of this model, both of which echo criticisms which also apply to simple abstraction. The first, which I have already discussed in Chapter 4, is that there is an element of presupposition in reflective abstraction: when the child notices that if $A<B$ then $B>A$, he uses the very scheme that is to be abstracted. As the preceding discussion has shown, Piaget attempts to forestall this situation by arguing that there is a qualitative difference between the original action scheme and the abstracted structure. In Chapter 4 I argued that Piaget's account of this difference was circular, while in the first section of this chapter I showed that, if there is in fact a qualitative difference between the two structures, then Piaget is left with a major problem of explanation with respect to the transition from weaker to stronger logics.

The second criticism, which is more central here, concerns Piaget's assumptions regarding the nature of action. For Piaget, the basis of the process of reflective abstraction is the structure of actions themselves: without underlying structures of action, reflective abstraction simply cannot occur, just as simple abstraction is impossible without recurrent essential features or qualities of objects such as 'redness' or 'squareness'. A major part of the argument against simple abstraction takes the form of an attack on the idea of abstractable essential properties of things, and a similar point can be made with respect to reflective abstraction.

Piaget's account relies on the assumption that any action can be described in terms of one underlying structure and, conversely, that actions which can be described in terms of the same formal structure bear a special relation to each other such that they require the same kind of understanding in their performance. But many actions, including those with which Piaget is primarily concerned, can be multiply described. So, for instance, the fact that a particular action can be described as being of a certain type (for example, putting pebbles into one–one correspondence) does not necessarily mean that the agent himself sees it in those terms, or will ever come to see it in those terms while he remains outside of the particular social context in which that action is so described.

Piaget, however, argues that because a child behaves in ways that in other contexts could be described as putting objects into one–one correspondence, or classifying, or relating them, then these are structures that he could, theoretically, abstract from his actions. But again, the abstraction of those particular concepts not only presupposes that the child possesses the concepts of one–one correspondence, classification, or seriation, but also relies on his participation in the relevant context. The fact that some actions can be described in many different ways means that the abstraction of, say, the concept of associativity from playing with pebbles presupposes some understanding of the context of doing arithmetic, if not the notion of associativity itself. Similarly, if a person is to understand an action as exemplifying associativity, that action must take place within an appropriate context in which it has that meaning, and not any other. Such understanding cannot be abstracted from the action by the individual alone; it involves entering into the social practices which surround that particular description of the action.

'The individual constructs his world': Piaget's account of objective knowledge

A central idea of Piaget's account of growth is the transition from weaker to stronger logics, by which he hopes to avoid the problems of empiricism and rationalism and steer a middle way between the two. The first two sections of this chapter criticized this notion from a variety of standpoints. In the first section I argued that there is no plausible way that a stronger logic can be generated from a weaker one since ideas expressible in the first cannot be reduced to the second. In the second section, I argued further that Piaget's notion of reflective abstraction cannot escape the problems of an ordinary abstraction model, since the abstraction of concepts such as associativity from the structure of actions presupposes possession of the concept in question. Not all actions can be adequately described in terms of one essential structure, and if a person is to understand an action as being of a particular kind, then that action must occur within a particular context.

These criticisms of Piaget's account suggest that what is required is a different kind of description of knowing about number and coming to know, which takes into account the social context of learning and the social practices involved in the use of numbers. In this section I will begin to develop this recharacterization of knowing about number by arguing that an account of growth must include social factors; in doing so I will draw on Hamlyn's (1978) argument that Piaget cannot explain the growth of objective knowledge. A reconceptualization of knowing about number as knowing how to act in particular social situations follows in Chapter 6.

The connection between equilibrium and objectivity in Piaget's theory is indicative of his general concern with the problem of how the individual constructs his world such that 'knowing reality means constructing systems of transformations that correspond more or less adequately, to reality' (1968a: 14). As such, Piaget's theory does not differ from the empiricist and rationalist philosophies which he criticizes, and which approach the problem of knowledge from a Cartesian point of view: the individual can be sure of his own existence, but not of anything else, the problem then being to assess the basis for the justification of knowledge claims about the relationship between individual experience and reality. Thus, as Hamlyn says of Chomsky (1968), Skinner (1957), and Piaget (1969, 1971b, 1971c, 1972a), they are all

> in that tradition in one way or another. They all take the epistemological problem as one that exists for the individual in relation to his world; it is not thought of as a problem which he must of necessity share with others if only for the reason that the terms in which the problem has to be expressed have a public and therefore inter-subjective meaning. Thus the individual is not thought of as one who shares a life, both cognitive and otherwise, with others, a fact which must inevitably colour the way in which the problem of cognitive development is construed.
>
> (1978: 55)

As Chapter 4 showed, Piaget's theory is very much concerned with delineating the principles by which objectivity is arrived at and what it involves; decentration, reversibility, and equilibration are key notions in this. With respect to the question of what objectivity entails, recall two earlier points I made in Chapters 2 and 4 about Kant's ideas: the first is that, for Kant, synthetic *a priori* propositions express the necessary conditions of the possibility of objective experience; the second is that Kant was not concerned with the problem of the *development* of objective knowledge, but with that of how we can claim to know anything. Now, Hamlyn argues, in the same sense that Kant merely states that the necessary condition for a judgement to be true for everyone is conformity to certain principles, but does not say how a person arrives at the point of being able to distinguish or know that they distinguish correctly objective judgement from subjective imagination, so Piaget's notions of decentration, reversibility, and equilibrium express the necessary conditions for objectivity but do not amount to sufficient conditions for the attainment of objectivity.[2] Both Piaget and Kant say what must be true for a judgement to be objective, but neither can account for the actual growth of that objectivity. Hamlyn argues that the Piagetian notion of decentration, for instance, does not explain the means by which growth towards objectivity and away from subjectivity is effected, but merely describes what must happen:

> That it does [imply a turning away from subjectivity] is true, but that fact implies only that we must find room for a distinction between what is due to or concerned with the self and what is not. It does not indicate how that distinction is to be given application. The notion of reversibility is not sufficient to provide the basis for the application of the distinction. To say that a form of thought involves reversibility is in effect to imply that it has a degree of logical sophistication, coherence, and rationality. However admirable it may be in many ways that a form of thought should have such qualities, they do not ensure that the form of thought is objective; what one thinks can be extremely logical without thereby being in the running for truth.
>
> (1978: 56)

It can be argued, then, that Piaget does not sufficiently explain how the individual distinguishes objective and subjective. With respect to his idea that cognitive development is a question of a progressive structuring of the world which follows a necessary path to maturity, Piaget fails to explain how by progressive equilibrations the individual arrives at knowledge that is in fact objective. Hamlyn argues that Piaget cannot explain by means of restructuring alone what having and acquiring a concept entails, because structuring refers merely to classifying and relating things without necessarily guaranteeing the objectivity of the principles of classification, and if having the concept is to count as knowledge, then those principles of classification must be objective. It is perfectly possible, then, to structure the world without that structure being an objective one. For Hamlyn, objectivity is social, and cannot come about through individual construction of the world:

> The acquisition of knowledge . . . is in effect the initiation into a body of knowledge that others share or might in principle share. This is because the standards of what counts as knowledge (and, since knowledge implies truth, the standards of what counts as true also) are interpersonal. The concepts of knowledge, truth and objectivity are social in the sense that they imply a framework of agreement on what counts as known true and objective.
>
> (1978: 57–9)[3]

The view that knowledge is intrinsically social entails that a psychological

77

account of knowing and coming to know can argue that, not only does knowledge imply agreement amongst people as to what is known, but that possessing knowledge will fruitfully be conceptualized as knowing how to act in a certain social context. Examination of Piaget's theory has shown that his account of the growth of knowledge is inadequate, and that his proposed mechanisms of reflective abstraction and equilibration fail to account for the growth from weaker to stronger logics. It may in fact be the case that, as long as the description of knowledge follows the same pattern of asocial formalization, an explanation of growth will be problematic or impossible. Thus in the first section of this chapter I argued that Piaget cannot demonstrate a plausible transition from a weaker to a stronger logic; in the second, that he cannot explain the link from action to cognition; and in the third, that he cannot explain how by progressive equilibrations the individual arrives at knowledge that is objective. The next step, then, is to argue for a reconceptualization of knowing about number, and this is the subject of Chapter 6.

6

Do number theorists give adequate accounts of knowing?

The purpose of this chapter is to examine what Piaget and other theorists mean by 'the concept of number', to consider the implications of their accounts, and to suggest an alternative description of knowing about number. My main criticism of Piaget's account is that it is 'essentialist', which is to say that it characterizes knowing about number in terms of the possession of certain essential principles that constitute the necessary and sufficient conditions for the possession of the concept of number; specifically, Piaget claims that the understanding of one–one correspondence is necessary and sufficient to indicate the number concept while the understanding of class inclusion and seriation are individually necessary and jointly sufficient conditions. Thus Piaget assumes that all the uses of number have a common factor, or essence, such that once someone possesses the essential principles of number they 'have' the number concept for all situations. An essentialist view supposes that meaning can be caught in an intensional definition, and thus assumes that the applications of number have a high common factor and are not merely related to one another. I will argue here that, in fact, the uses of number may be similar, or related, but that there are important differences between them which cannot be captured by a common factor.

In earlier chapters I considered the problem of how to account for development, and argued that an account of knowing such as Piaget's has inherent problems in this respect. I suggested that an adequate account of growth must include social factors, and further, that knowing about number should also be reconceptualized as intrinsically social. This chapter arrives at the same conclusions via a slightly different route which takes the form of a criticism of Piaget's essentialism. If essentialism is rejected in favour of the view that the many different applications of number are not united by a common principle but on the contrary differ, such that knowing about number is a question of knowing how to act in different situations, then, in so far as these different situations are social situations given meaning by the community, learning about number is a question of entering various social practices. Thus knowing about number can be described as intrinsically social; in the first section of this chapter a criticism of Piaget's essentialism leads to a reconceptualization of number understanding and its development. The second and third sections continue this argument with an analysis of the concept of number as described by Bryant and Gelman and Gallistel; here the aims are to identify essentialism in these accounts and to question their adequacy. This chapter also begins to analyse the context of various number uses and to examine the relationship between these and the contexts of psychological experiments.

Piaget's essentialism

In this section I shall argue that (i) Piaget gives an account which I shall call 'essentialist', by which I mean that he characterizes knowing something as knowing certain essential principles or as necessitating a particular structure of thought; and (ii) Piaget's account of knowing is not the only one possible, and can be replaced by one which reconceptualizes number understanding in terms of entering into social practices.

To recap on Piaget's account of what it is to know number, which I initially introduced in Chapter 3:

(i) Piaget does not consider the ability to count to be sufficient indication of having the number concept. Instead, the criterion he sets for this is the ability to conserve number: that is, the recognition that if two sets are in perceptual one–one correspondence, the movement of one set so as to destroy the perceptual correspondence does not entail that the two sets are no longer equal.

(ii) According to Piaget, the concept of number is formed from the synthesis of two groupings: classification and seriation. Piaget's description of the development of the number concept constitutes an exception to the rest of his theory; usually he is concerned with the general logical structure of thinking at various stages of development, such that the development of concepts such as space and time are only described as manifestations of a general structure of thought at any particular stage. Number constitutes an exception, not only because it is treated as a discrete concept, the development of which illustrates the development of the overall structure of thought, but also because the synthesis of two groupings that underlies the number concept is unique to it, and does not happen in thought generally. During the formation of the number concept, the two types of reversibility (reciprocity and inversion) are synthesized, but this does not happen on a general level until the formal operational stage. It is not until this stage that reciprocity and inversion are combined so that both are readily available. Thus it is only with respect to the number concept that the synthesis of reciprocity and inversion occurs at the concrete-operational level, and it occurs as a special case.

Thus possession of the concept of number relies on and manifests a certain structure of thought. In particular, this structure is illustrated by the ability to conserve number.

(iii) Knowing number entails: (a) understanding one–one correspondence; and (b) understanding class inclusion and seriation. Piaget's characterization lists (a) as an essential principle (it is necessary and sufficient to indicate a grasp of the number concept) and (b) as logical prerequisites for understanding number (individually necessary and jointly sufficient conditions). Underlying these is the concrete operational structure of thought, of which these are exemplars. Knowing number, according to Piaget, by definition requires that thought exhibit this particular structure, and be capable therefore of a particular form of logic which is stronger than that of the stage preceding it and weaker than that of the stage which follows. To clarify the position: the child does not learn this logic *per se*, but learns new concepts which are indicative of a progressive restructuring of thought. That is, her capacity for solving certain problems or apprehending certain facts about the physical world develops as the structure of thought becomes more complex. To use Fodor's phrase, the expressive power of the child's logic develops. Put this way, learning new concepts can be the product of development, but what changes is the underlying structure of thought, which is not itself learnt. Thought becomes more developed through action in the world.

It follows, then, that Piaget's account of having the number concept explicitly requires that children understand class inclusion, seriation, and one–one correspondence. These must be understood in the sense that a child must be able to perform tasks that involve these constructions. An appeal to language difficulties as a possible barrier to their successful execution is unsatisfactory according to Piaget, whose stance on language is that it follows after and hence merely reflects the development of thought. Essential to the understanding of number are the transitive-asymmetrical relation, class inclusion, and one–one correspondence. According to Piaget these indicate the necessary underlying logical structure for having the number concept. This can then be called an essentialist account of knowing number.[1]

Criticism of essentialism and an alternative account of knowing number

Piaget's account of number is not the only one possible, however, and it is not only possible, but also necessary, to give another sort of account of knowing, because an essentialist account poses serious problems for any attempt to explain how someone can learn a new concept. It is necessary to explain how it is possible to develop from one level of thought to another level which is richer and includes structures which were missing from the first; Fodor argues on *a priori* grounds that this is not possible. Piaget's account of growth relies on the notion of reflective abstraction to explain the progress from weaker to stronger logics, but as a complex form of ordinary abstraction, reflective abstraction does not escape the problems of abstractionist accounts.

In particular, abstractionism assumes that having concepts involves the apprehension of recurrent, essential features of objects or groups of objects, and that these features can be abstracted from the environment by the individual. Abstractionism features in a number of other accounts, for instance Klahr and Wallace's (1976) and Riley *et al*.'s (1983) information-processing accounts, and also in Brainerd's (1979) 'ordinal theory' of number. Its main problem is that learning by abstraction presupposes the very concept to be abstracted: one cannot abstract the concept 'red' from the unstructured environment without knowing something of the concept 'red' before, or without having a particular structure of thought. This argument against abstraction coincides with Fodor's argument, which states that it is impossible to generate a stronger logic from a weaker one, and that if a concept belonging to a stronger logic has prerequisites which are expressible in a weaker logic then either the concept does not truly belong to the stronger logic or there is no real distinction between the two logics. Any insistence on the existence of different qualities of logic leads to the conclusion that it is impossible to develop from the weaker to the stronger.

Another way of describing concept learning is to say that having a concept does not entail having certain essential logical prerequisites, but on the contrary entails having an understanding of how to act in certain situations. This would mean that there is no underlying structure, logic or set of principles to be developed. Instead, what is required is an understanding of social situations and what they mean to other people, and thus an entering into social practices. A consequence of this is that acts which appear to be the same in formal terms may not in fact be the same with regard to what understanding is required to perform them successfully and appropriately. The formal structure is not important because it does not feature in the actor's understanding, which refers to contexts as a whole rather than to isolated acts. According to this argument, knowing is intrinsically contextual; this means that knowing contains a social component, an idea that I will illustrate in Chapter 8. In this section I will indicate further why an essentialist account of knowing number is inadequate.

An initial proviso is that to argue against an essentialist account on the grounds that it is inadequate to explain the understanding of number does not lead to the conclusion that it is therefore impossible to say what must be true of someone if they are properly to be said to know something. An argument against an essentialist account of number does not rule out an account which says that in order to perform successfully in a number-conservation task a person must understand what is required from them in this situation, in what way certain words are being used, what a test of this kind is for, and so on. This would not be an essentialist account whereas one which said that an understanding of one–one correspondence is necessary and sufficient for the proper use of number would be.

It can be argued that apart from the fact that an essentialist account poses problems in explaining the growth of knowing in terms of essential principles, such an account is inadequate as it stands simply in terms of describing knowing. Wittgenstein (1953) attacks essentialism on the grounds that the 'commonsense' belief in the essences of things is mistaken. Rather than looking for similarities, Wittgenstein looks for differences, arguing that when the various individuals to which a given general term applies are examined, there is nothing to be found which they all have in common. According to Wittgenstein, it is a matter of fact that they do not share a common essence. His famous example concerns games, where, he argues, there is not something common to all games, but rather a whole series of similarities and relationships. This, however, is not to be taken to mean that we are dealing with a motley, disconnected group of things arbitrarily called by the same name; Wittgenstein denies essences, but not 'family resemblances':

> we see a complicated network of similarities overlapping and criss-crossing: sometimes overall similarities, sometimes similarities of detail . . . I can think of no better expression to characterise these similarities than 'family resemblances'; for the various resemblances between members of a family: build, features, colours of eyes, gait, temperament, etc., etc., overlap and criss-cross in the same way. And I shall say: 'games' form a family.
>
> (1953: sections 66–7)

And concerning number, Wittgenstein says:

> the kinds of number form a family in the same way. Why do we call something a number? Well, perhaps because it has a – direct – relationship with several things that have hitherto been called number; and this can be said to give it an indirect relationship to other things we call [by] the same name. And we extend our concept of number as in spinning a thread we twist fibre on fibre. And the strength of the thread does not reside in the fact that some one fibre runs through its whole length, but in the overlapping of many fibres.
>
> (1953: sections 66–7)

Following Wittgenstein, it can be argued that 'no general term has a unitary meaning' (Pitcher's commentary, 1964: 219). According to Pitcher, 'a word has a unitary meaning when its meaning refers to certain definite characteristics and something must have all of them for the word to be properly applicable to it' (219). So, to take the games example, suppose that there is a group of characteristics $C_1 \ldots \ldots C_n$ which typify games. Pitcher argues that if all games had all of these characteristics, and only games did, then the word 'game' would have a unitary meaning. However, this is not the case, so that to be a game, some of the characteristics must apply but not necessarily all; also, some combinations will not do, while others, which may contain fewer characteristics, will. The main point is that it is impossible to specify in abstract the necessary and sufficient conditions for being a game. This general argument can be applied also to the seemingly more specific term 'lemon'; the characteristics which typify lemons can be listed and we can say that if an object has all of these it is certainly a lemon. But, Pitcher points out, if there occurred a strain of pink, sweet, lemons, these would still be lemons of a special kind. While it is not possible for a thing to lack too large a number of the properties which typify lemons, it is on the other hand impossible to lay down any particular property or groups of properties which a thing must have in order to be a lemon.

Following this argument through, it is necessary to reconsider the question of what is involved in knowing that something is a lemon, or a game, or a number. The question concerns the grounds on which knowledge of this kind can be attributed to someone. Wittgenstein first observes that to speak a language involves many different skills and abilities, including non-linguistic behaviour. To take Pitcher's (1964) example, a child learns the meaning of a new word 'ball': we do not attribute this new learning to a child if she is merely able to say the word 'ball' like a parrot, but neither should we attribute it to her if she has learned to reply 'ball' if someone, pointing to a ball, says, 'What is this?'. This is because an ostensive definition does not carry a full interpretation of a word. The second part of Wittgenstein's argument shows why this is the case. An ostensive definition of the word 'ball' involves pointing to a ball and saying, 'This is a ball'. But how much does this convey? Pitcher says that

> it is natural to suppose that such a definition uniquely determines the meaning of the word 'ball', and hence that the child, in being able to repeat the manoeuvre, must know what that meaning is. But Wittgenstein shows that this supposition is false. In pointing to a ball one is at the same time pointing to a round thing, to a thing of a certain colour (e.g. red), to a thing of a certain size, to a thing of a certain weight, to a thing belonging to a certain person (e.g. Johnny), to one thing, to a thing made of a certain material (e.g. rubber), and so on. Hence, the ostensive definition, by itself, does not uniquely determine the meaning of the word 'ball', and the child, in repeating it, does not necessarily know what that meaning is . . . . [Ostensive definitions] do not in

themselves guarantee success, for they must in every case be properly *construed*, properly *interpreted*, properly understood.

(1964: 241)

If an ostensive definition does not convey the proper meaning of a word, then how can a child be judged to have learnt what 'ball' means? Wittgenstein argues that this judgement can only be made by observing certain sorts of behaviour in the child. Thus a child may be able to fetch a ball, draw one, pick one out from a number of objects, speak about one appropriately, and so on. This behaviour may give a reasonable indication that the child knows what 'ball' means. But:

> there is no specific point after which it will be absolutely certain, without any possibility of being proven wrong in the future, that the child knows the meaning of the word 'ball'. But after observing his behaviour in a variety of situations over a certain period of time, we can be reasonably certain that he does.
>
> (Pitcher 1964: 242)

Thus speaking a language involves certain behaviours, abilities, and skills such that Wittgenstein says: 'I shall also call the whole, consisting of language and the actions into which it is woven, the "language-game"' (1953: section 7). Thus the meaning of a word is its role in the language-games in which it plays a part, so that a word is embedded in the kind of behaviour that surrounds its use. Words cannot be treated in isolation from the actual practical situations in which they are used. Thus Wittgenstein says: 'But what is the meaning of the word "five"? No such thing was in question here, only how the word "five" is used' (1953: section 9).

In contrast with these ideas, Piaget's theory is essentialist, in that it assumes that having a concept of number entails the possession of certain essential principles. Piaget's claim is that if someone does not know the essential principles of number conservation, class inclusion, and seriation, then they do not know number. He is not satisfied with someone's being able to count things, and up to a point it might be possible to agree with this position, but for different reasons: namely, on the grounds that this shows that someone only knows how to use numbers in one way. As with the example of knowing the meaning of 'ball', I will argue that understanding of the number concept cannot be confidently attributed to someone until they can do more with number than counting things. In order to make a fairly confident judgement that someone knows the meaning of numbers one might require not only that they can count things, but are also able to measure things with rulers, use money, follow a recipe, share things out, follow numbered instructions, play cards, catch buses, and find an address in New York, for instance.

To follow Wittgenstein's argument, it could be said that if a person can do all these things they understand the meaning of numbers, but it is not possible to set a minimum criterion for what understanding of numbers requires, which is what Piaget wants to do. Nor can it be said that underlying these uses of number is an essential principle or principles which enable the person to perform the acts in question and which we simply need to find evidence of in someone's understanding in order to be able to attribute possession of the number concept to them. If the meaning of number words is embedded in the behaviour which surrounds their use, then, in the same way that it is not possible to set a minimum criterion for having the number concept, it is not possible to say with absolute confidence that if a person can do all the above things, they can then understand the meaning of numbers. But it is possible to be reasonably certain.

Reconceptualizing number development

Part of the argument against an essentialist account is that there are problems in the idea that learning a concept proceeds by ostensive definition or abstraction. An alternative account of knowing number as knowing how and when to use numbers not only makes better sense as an account of knowing, but it also generates a more coherent explanation of how a child comes to know. I shall argue here that, in so far as the meaning of a word is constituted by its use, and in so far as speaking a language is to engage in 'forms of life' which provide the foundation for 'agreement in judgements' (Wittgenstein 1953: 1, 242), then knowledge is intrinsically social, while knowledge gain is a question of entering into a social existence.

For a child to enter into a common social understanding, she must, in the case of number, be able to use and respond to numbers as others do. In this sense, the child who counts '1, 2, 4, 3, . . .' does not know how to count, because this is not how others count. In order for a person to come to do things as others do, they must be aware of the responses of others to things they say or do, and specifically, must know what it is to be in agreement with others and to share their judgements and expectations. To take an example: suppose a child counts '1, 2, 4, 3, . . .' and is corrected by someone else, who says, 'No, it's 1, 2, 3, 4'; if the child changes her way of counting in accordance with this correction, it may be said that she knows that she now counts in the right way because someone to whom she attributes greater knowledge of these matters says that that is the right way to do it. Knowing the correct way to count presupposes knowing how to act in a particular situation, which in turn presupposes agreement between people. So knowing what it is to be in agreement with others, and sharing their judgements and expectations is what constitutes knowing what it is for something to be true or correct.

In changing the way in which she counts, the child shows that she appreciates the force of the norm, to use Hamlyn's (1983) phrase.[2] Entering into social practices does not involve mere awareness of whether others agree or not; coming to know in this sense also involves a *response* to the force of a norm. This response is also dependent on a social existence: coming to act according to the norms of appropriateness can be seen as a question of the authority of 'significant others' within a context of personal relations. So, to return to my example of the child who counts wrongly and is then corrected, the child might take the adult's version to be correct because that person is a 'significant other' with more power in that they hold authority on how things are to be counted. It is only on such a basis that the child is to choose between her own and the adult's version as the correct one. This is not to say, however, that the child has a concept of correction or truth, but it can be said that she knows what it is for something to be true in so far as she is aware that (and cares whether) significant others agree or do not agree with statements she makes, accept or correct actions by which she responds to commands and questions, and act or do not act according to her expectations. Hamlyn (1978) describes such a context of personal relations as follows:

> The appreciation of rules . . . is likely to emerge out of an appreciation of 'what *they* want'. I mean that the child has to come to accept the existence of other interests, wants, wishes, feelings and attitudes, sometimes agreeing but often not, with his own . . . . The treatment of the child progressively as a person, and therefore as both the recipient and source of human feelings and attitudes, is of vital importance in the child's becoming a person in a real sense . . . . Hence emotion is not simply a distracting and irrational factor . . . an essential stage in the acquisition of knowledge of what it is for something to be X is, through this: distinguishing Xs 'as *they* do'.
>
> (1978: 101–2)

Thus a reconceptualization of number understanding and its development in terms of entering into the social practices of number use demands recognition of the importance of a social existence in knowing and coming to know. On this basis I will argue that Piaget cannot give an adequate account of knowing about number: his essentialist approach leads him to search for the possession of logical concepts representing the necessary and sufficient conditions for knowing number, on the assumption that tasks which are formally isomorphic to those concepts are solvable if a child has the number concept. He thus overlooks the possibility that possession of those concepts is not what is actually required to solve the tasks, and that knowing about number does not necessarily have to be shown by the ability to solve them.

Reconsidering Piaget

So what is the relevance of being able to perform successfully in Piagetian tasks? Piaget considers number conservation, class inclusion, and seriation to be the logical and psychological roots of the number concept for the reasons set out in earlier chapters. But what do the associated tasks have to do with understanding number use, and what kind of understanding does success in the tasks signify?

Taking the conservation of number test first, a number of commentators (e.g. Rose and Blank 1974; McGarrigle and Donaldson 1974) have remarked that success in the test does not rely on number understanding but, rather, on the correct negotiation of potentially misleading social cues. Many other interpretations of the conservation-test situation are possible besides the one that Piaget intends, and the test may have more to do with understanding the use of the words 'more than', 'less than', and 'the same as' in a particular context than with understanding one–one correspondence or numbers; I will pursue this point in the following chapters.

Similarly, the class-inclusion test assumes that if a person has the number concept they will also (by definition) know that, if there are six black cows and four white cows, then there are more cows than black cows because 'black cows' is logically included within the superordinate class of 'cows', and also because the number six is logically included within the number ten. So, Piaget says, failure to correctly answer the class-inclusion question, 'Are there more cows or more black cows?' indicates a lack of understanding of one of the necessary conditions for having the number concept. The test can, however, be reinterpreted and the argument put forward that its correct solution does not tap number understanding at all but, rather, requires the appropriate interpretation of the word 'more' in this particular situation. To reanalyse the test: there are ten cows, six black and four white, and someone asks if there are more cows or more black cows. The usual adult reaction to this is that it is a trick question, and it requires more thought than usual. This simple fact cannot be wholly without significance, and leads to the conclusion that when a child does get to the stage of being able to solve class-inclusion problems, it is because she has understood the situation not in terms of black and white cows but in terms of being asked that kind of question – a question which is rather like a trick or a puzzle. A similar situation is one in which someone is asked: 'Which is heavier, a ton of feathers or a ton of bricks?'. When people answer this question wrongly, it is not assumed that they lack a proper concept of mass or weight, but rather that they have been successfully tricked into assuming that the conventional use of the word 'heavier' is being used – that is, the use which compares two different weights, not the same one. When someone asks if something is heavier than something else, they do not normally put the answer in the question, unless they are playing a trick and are not in fact making a genuine request for information. So this is a special use of the word 'heavier', and in the class-inclusion test there is a special use of the word 'more'.

Finally, seriation tests do not refer to number at all. Piaget argues that seriation and number are connected as a point of logic, because seriation can be formally represented in terms of the transitive-asymmetrical relations which he thinks underlie the number concept. But here again success in the task may simply be a function of a child's interpretation of the overall situation rather than a question of making deductive inferences: it is quite conceivable that a child could think that the inference question concerning the relative lengths of two indirectly compared quantities is unanswerable, but at the same time is able to use a ruler to measure things, not because she understands – or needs to understand – the logic of transitive-asymmetrical relations, but because she understands the workings of this particular social practice. Indeed, it is doubtful that anyone who uses a ruler stops to think why this is a reliable method of comparison.

Whatever the reasons, though, for failure in tests designed to tap the logic of number conservation, class inclusion, and transitive-asymmetrical relations, the main argument here is that Piaget's essentialist position leads him into an unfounded assumption that ability to do these three tests constitutes the necessary and sufficient conditions for understanding numbers. On the contrary, the tests may have little to do with the use of numbers, while knowing about numbers involves a series of skills which are to do with using number words in various contexts to achieve various ends.

If understanding numbers involves knowing a series of applications of number in different situations, this knowledge can be shown in a variety of ways, through a variety of behaviours. Coming to know can be described as an initiation into various social practices in which learning is characterized by a movement towards participation in the common understanding of what numbers stand for and do. This means that children's interpretations of Piagetian tasks, or other tasks to do with numbers, are not merely defective or incomplete when they differ from the adult shared view; rather, they represent rational ways of dealing with the world given the fact that children have limited social experience.

Bryant's work

In this section, I aim to show that the criticisms I made of Piaget's theory in the previous section are also applicable to Bryant's (1972, 1974) work. Although Bryant is highly critical of Piaget's account, examination of his theory shows that it does not differ substantially from Piaget's. In opposition to Piaget, Bryant aims to show that children can have access to the principle of number invariance and still fail conservation of number tests. However, like Piaget, Bryant assumes that: (i) having the number concept entails having access to essential logical principles which are necessary and sufficient to generate the number concept: in particular he assumes that if a child has the concept of number then this can be assessed by any test isomorphic in structure to such principles. In so doing, Bryant also assumes with Piaget that: (ii) the part played by context in knowledge is unimportant: he assumes that, in any task isomorphic to a particular logical structure, the problem can be solved solely by reference to that structure, and that no other information or understanding is necessary. Although he considers it possible for logical competence to be blocked by the misuse of certain conceptual tools, Bryant does not consider the possibility that understanding of the situation itself is in fact intrinsic to success in the task in question.

Thus Bryant, like Piaget, ignores the problems associated with an essentialist view, namely the objection that conditions for knowing the meaning of a term cannot be specified *a priori*, because the meaning of a term such as 'number' is embedded in the behaviour that surrounds its use. Both accounts, then, 'decontextualize' knowledge, overlooking the view that the assessment of a child's understanding of numbers requires analysis of how far a child can be said to have entered into the social practices of number use.

Bryant's theory

For Bryant, number can be apprehended as an objective quality of the environment in the same way as orientation and size can be; unlike these two perceptual continua, however, number is not as amenable to coding in terms of the relationship between things. The orientation of a line, for instance, can be fairly efficiently represented in terms of its relation to another line; similarly, if a judgement is required as to which of two sticks in a pair is the longer, then a relative judgement will do as well as an absolute judgement in terms of the length in inches of each stick. Again, the comparative numerosity of two rows of counters can be coded in the relative terms of length, one–one correspondence, or density, but only the relative code of one–one correspondence is accurate, while length and density are frequently misleading; where one–one correspondence cannot be used, the absolute code – numerals – is best. According to Bryant, the errors children make in number-conservation tests are due to the fact that they use conflicting relative codes for number, not to the fact that they do not understand the principle of invariance.

Perceptual cues and the relative number code

Bryant begins, then, with the assumption that young children use relative

90

codes for number, as well as for other features of the perceptual environment such as size and orientation; it is only when they are older that they begin to use 'internal frameworks' or absolute codes – for instance the number system in the case of numerosity. Clearly, the use of relative codes can be disadvantageous: it is difficult to record information about individual stimuli since this would have to take the form of a relation to a background feature if one exists, and this information would then have to be used inferentially, in order to compare individual stimuli seen at different times. This disadvantage is increased in the case of relative number codes: the use of the relative code in number does not tell a child anything about the actual number of a group of things, and, unlike perceptual continua, there is no external frame of reference to use to overcome its weaknesses. If is for this reason, Bryant argues, that 'the young child's codes for number are initially a great deal less effective than his equivalent codes for perceptual continua' and, furthermore, 'some of the cues which he uses to make relative judgements about number are quite wrong. As a result his immediate relative judgements can often be totally incorrect' (1974: 113–14).

There are three relative codes for number that children might use: length, density, and one–one correspondence. The first of these, length, appears to be used by children when they judge a row of counters to be 'more' than another, shorter row, even though the longer one may contain fewer counters. Bryant argues that children do not say that the longer row has more simply because they have misinterpreted the experimenter's question; he reasons that they are indeed making a number judgement because sometimes the same children use the other, correct, cue of one–one correspondence to make number comparisons. Furthermore, Bryant argues, these children do not appear to note the difference between the two cues: they treat them as interchangeable and equivalent, using each as consistently as the other, depending on which cue is most salient in the display (Bryant 1972). Thus Bryant argues that:

> It is often maintained that young children are not really responding to number differences until they know their numbers. This is quite wrong. When the child who cannot count uses the one–one correspondence cue he is responding to a number difference consistently and correctly, but he is responding in a relative way. The fact that children use this cue before they can count is a very clear indication that the development is likely to be from relative to absolute codes.
>
> (1974: 116)

The problem with this claim is that Bryant has not proved that, given two rows of counters in various display formats, children are in fact making *numerical* judgements: he makes an inadequate distinction between considerations of length and matching and the use of numbers in the 'Which is more?'

task. The notion of using length as a cue to make number judgements presupposes that children might have a conception of number which is completely wrong. But when a child responds in terms of length, there is no evidence to show that she is making a number judgement, rather than that she is interpreting the word 'more' in some other way to make a reference to length. Because, as Bryant points out, a child apparently understands the word 'more' to be a reference to number in some situations (and this does not necessarily have to be the case even when the rows are in one–one correspondence, since length can still be the referent with unequal rows), this does not mean that she will understand it to be a reference to number in all situations. Since the word 'more' is in fact used in a variety of ways, there is no reason to expect that a child will know which way is intended every time, especially, as I will show in Chapter 8, in situations which are unfamiliar and ambiguous.

Bryant's treatment of number is one which assumes that it is possible to talk sensibly about number as an abstract quality or thing which can be separated from its use. This is the only way to make sense of his notion of absolute and relative codes; these are treated as ways of accessing an objective quality which is number, rather than as judgements of length and matching as well as of number. Bryant's treatment, particularly his idea of codes (but codes for what?) amounts to an approach in which it is assumed that children know about number, or can see it as though it were a quality like colour, but have inadequate ways of accessing it.

The invariance principle and conflicting relative code cues

Like Piaget, Bryant takes the ability to conserve number as the criterion for being said to have the number concept. However, Bryant believes that the Piagetian test for conservation results in an underestimation of a child's understanding of the principle of invariance. According to Bryant's theory, the correct conservation response can be arrived at by using either a relative or an absolute code: use of the absolute code can enable the child to solve the problem by simply noting that the numbers of objects in each row are the same; alternatively, use of the relative code enables her to reach the correct solution by judging that: (i) the two rows are equal because they are arranged in one–one correspondence; (ii) by the principle of invariance, the transformed row is the same before and after; and so (iii) by inference, the non-transformed row must equal the transformed row. Both (ii) and (iii) are necessary to this process because, once the perceptual one–one correspondence is destroyed, there is no relative cue to give the correct answer. Bryant argues, then, that:

This is a deductive, transitive inference, and it is true that if the child
is using only a relative code and manages to solve the conservation
problem, he must understand invariance, and he must also be able to
make a deductive inference.

(1974: 129)

Dismissing Piaget's explanation for failure on the conservation test, that
is, that non-conservers treat perceptual change as a real change (do not
understand invariance), and also the alternative possibilities that they either
forget that before the transformation both rows looked the same, or fail to
make the necessary deductive inferences, Bryant argues that children fail
Piaget's task because of a conflict between incompatible judgements.
Memory, he argues, is not a problem in this type of test: both Bruner *et al.*
(1966) and Bryant (1972) have shown that children can remember the
appearance of a display before transformation; similarly Bryant and Trabasso
(1971) demonstrated that children can make inferences in the context of a
transitivity task.[3] Thus, working on the premise that children use the one–
one correspondence and length cues interchangeably to judge the relative
manyness of two rows, and therefore consider both to be equally usable cues,
Bryant argues that the child who changes her judgement in the number
conservation test does not necessarily think that the rows have been changed
in number. Rather, she has used two types of cue which produce conflicting
judgements. A child in this situation has no means of distinguishing between
judgements, considering both equally good, because they are based on
equally good cues. She thus has no means of choosing one over the other.
According to Bryant, she may understand that the perceptual transformation
does not actually change the numerosity of the array, but nevertheless will
be unable to resolve the conflict.

Thus Bryant's explanation for failure on the test of conservation centres
on the child's use of an inadequate relative code: the one–one correspondence
cue is correct, but the length cue is not. Invariance is independent of this;
a child can understand invariance but still not perform successfully on the
conservation task because of her use of the inadequate relative length code.
Thus use of a relative code, unless the user knows that only one–one
correspondence is efficient, will half of the time lead to failure, and unless
conflict is removed by discarding the inadequate length cue, a child will be
unable to take note of the invariance of the transformed row.

But Bryant's explanation is bizarre, since an idea of invariance (nothing
added or taken away) is all that is necessary for conservation. There is no
need to judge the relative numerosities in this case, and possibly it could be
said that an understanding of invariance presupposes precisely this observation: that enumeration – whether 'relative' or 'absolute' – is irrelevant.
Furthermore, looking back at Bryant's three-step 'recipe' for correct relative
code solution of the conservation task, we see that there is no mention of a

post-transformation numerosity judgement – indeed, if one understands invariance, it is not necessary. And what is more, in the standard task, the rows are set up in an initial one–one correspondence which, according to Bryant's observations, will lead the child to make a one–one correspondence-based judgement – a first step in the direction of a correct answer. Bryant's claim is that the conservation test can result in failure even for a child who understands invariance, but the idea that this understanding can be blocked or overridden by the conflict between two codes is highly questionable. Nevertheless, Bryant favours the conflict hypothesis and pursues it experimentally, concentrating on the predictions that: (i) if conflict is removed, the child can apply the invariance principle; and (ii) showing the child that one of her judgements is not sound will resolve the conflict.

The first of the predictions Bryant (1972) demonstrates in an experiment featuring the displays shown in Figure 6.1. Each child was shown one of the A displays, which was then transformed into the C display. Bryant hypothesized that because the C display removes conflict by taking away the two relative code cues, an otherwise non-conserving child should be enabled to put the invariance principle into operation: her performance on C should

Figure 6.1 Displays Used to Test Bryant's (1972, 1974) Conflict Hypothesis

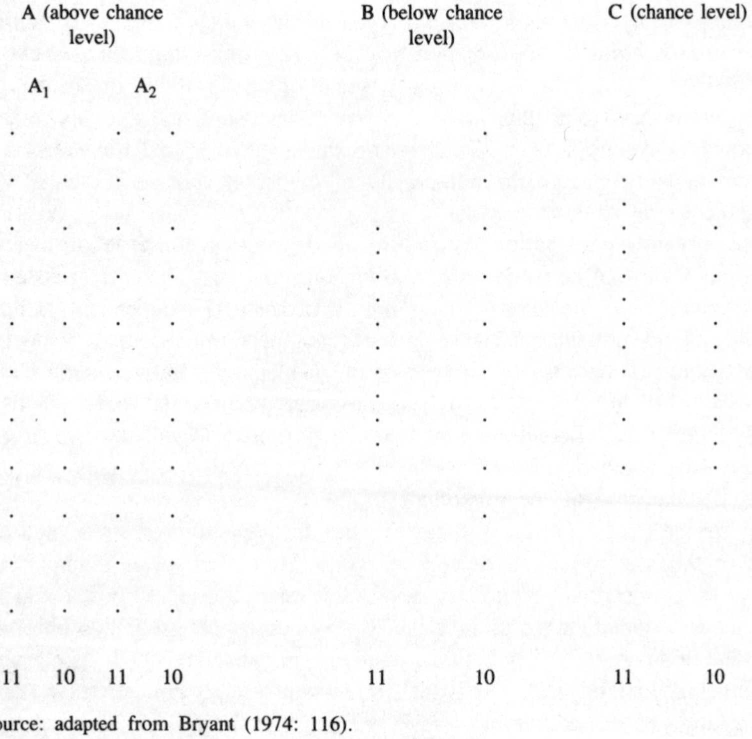

A (above chance level)		B (below chance level)	C (chance level)
A_1	A_2		

11	10	11	10	11	10	11	10

Source: adapted from Bryant (1974: 116).

reflect her understanding that the number of counters in each row has stayed the same, resulting in a better than chance level of accuracy in judging the relative numerosity of the two rows in the C display (shown the C display on its own, children tend to make judgements at chance level of accuracy, compared to an above chance level for A displays and below chance for the B display).

Bryant did indeed find this result, but there is an alternative explanation for it. Note that, ordinarily, C is at chance level; presumably this is because both rows look the same, but the situation demands an answer to the question, 'Which is more?', so children will produce an answer which is merely a guess. If performance on C improves after transfer from A, this may simply mean that the children are being conservative: there is no obvious answer for the C display, so they just repeat their previous answer.

Taking the conflict hypothesis further, Bryant also hypothesized that if children think that both length and one–one correspondence judgements are equally sound, then they should also transfer information from B when there is no conflict with the second display. Thus it was expected that incorrect judgements should be transferred from B to C (no conflict), resulting in below chance performance, but not from B to A (conflict between length and one–one correspondence), which should stay at above chance level. Again, the results were as expected, but again the same objections apply: transfer from B to C could in fact be the result of conservatism, while the stability of the above chance performance for A does not necessarily denote the use of invariance, but, rather, a consistency on the part of the child in using a matching procedure to answer the 'Which is more?' question in response to the A display. Again, the objection can be raised that a grasp of the idea of invariance precludes the necessity of making a relative numerosity judgement after the transformation.

Bryant's second prediction from the conflict hypothesis is that, if a child is shown the unreliability of the length cue, then conflict will be resolved. This effect can be achieved by training children to recognize when their relative judgements are sound or unsound: for instance Gelman's (1969) training experiment simply involved giving children feedback on their judgements in tasks requiring them to pick the odd one out of groups of three quantities. The children were trained not to respond to the length cue and in fact proceeded to do better on the ordinary number-conservation test. Bryant comments that in this experiment the children were not shown why length was a bad cue; in real life, he says, they would abandon the cue for a reason, possibly that the judgements prompted by it are inconsistent over time. Bryant accordingly devised an experiment to show children that the one–one correspondence cue is reliable while the length cue is not. Children were shown either A displays or B displays as in Figure 6.2. In the B display condition, which encouraged use of the length cue, the children watched while one display was transformed into another a number of times; each time

Figure 6.2 Displays Used to Demonstrate to 'Non-conservers' the Unreliability of the Length Cue versus the Reliability of One–One Correspondence

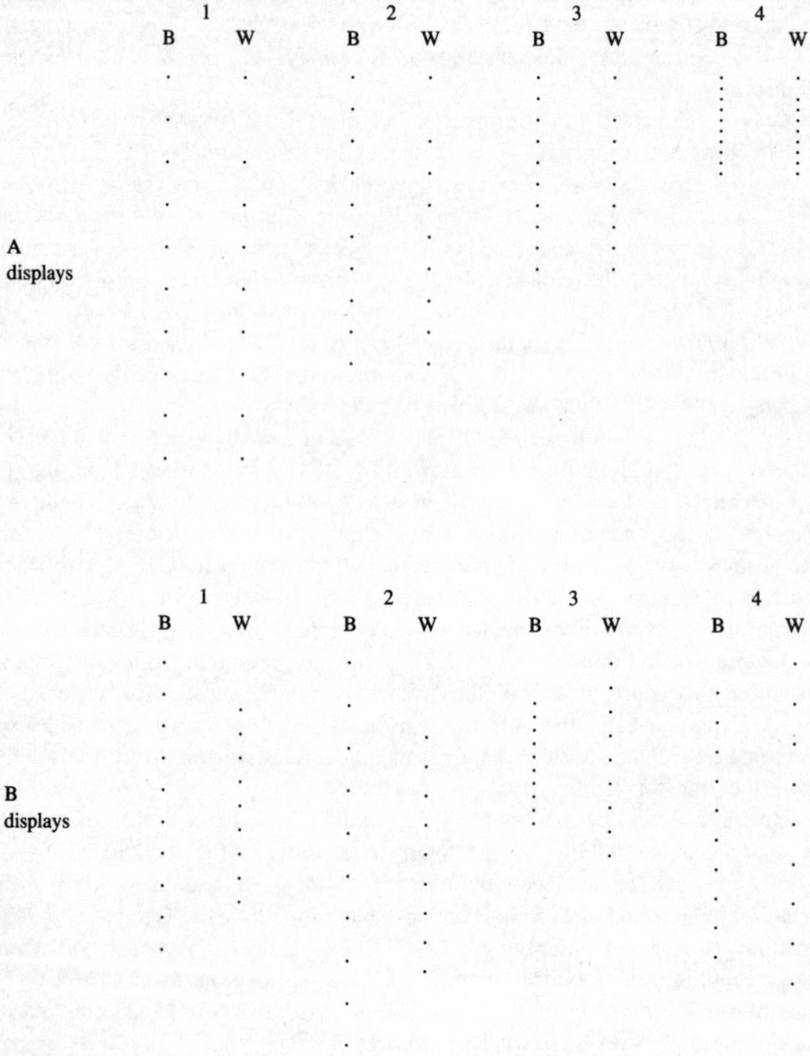

A
displays

B
displays

B = black
W = white

Source: Bryant (1974: 143).

the children were asked which of the two rows had more counters. Their use of the length cue thus led to inconsistencies, the black row being the more numerous half of the time, and the white row being more numerous the other half. In the A display condition, which encouraged the use of the one–one correspondence cue, the same transformations did not lead to inconsistencies: the black row was always the more numerous. The training sessions were preceded and followed by the two conflict transformations A–B and B–A from the previous experiments.

The results were as Bryant predicted: instead of abandoning the one–one correspondence cue in favour of the length cue in the A–B conflict transformation, the children continued to use one–one correspondence. Thus, Bryant argues, the children must have an understanding of invariance since, following the experience of making continually changing judgements with the B displays, they change their conservation judgements. Once again, however, it is arguable that this experiment does not show a change in the child's faith in length as a code for numerosity but, rather, a change in her understanding of the aims of the conservation-type exercise – that is, that a judgement in terms of number is indicated by the use of the word 'more', rather than a judgement in terms of length. Indeed, since it may be the case that experience with the conservation task leads to improvement simply because the experimenter's intentions become clearer, it may be that Bryant's training procedure is redundant. (It is significant that the effect of experience in Bryant's experiment was not measured: the control group were not simply exposed to the ordinary task, but instead received training sessions involving only B displays in which the length cue always led to the same judgement – that is, the row with 'more' was always the same colour. The result of this would be merely to compound the child's interpretation of the task as one of judgement of length, with the unsurprising result that the same interpretation is made in the post-training task.)

Bryant's work: summary

Bryant's major concern is to introduce adequate controls into the conservation task so that a child's performance reflects her understanding of invariance without being blocked by a misuse of relative codes for number. Thus Bryant considers his version of number conservation to be a 'purer test of the understanding of invariance' (1974: 175) than Piaget's original test. But in explaining children's non-conservation behaviour, Bryant never establishes whether or not they do in fact interpret the experimental situation to mean that they are supposed to be making numerical judgements, as opposed to length judgements. He assumes that children are making some sort of numerical judgement and using length as a code for this, and so never takes a child's length judgement at face value or asks why she is making such a judgement when the experimenter's intention is that she makes a numerical judgement. For Bryant, the child's understanding of the experimental context is unimportant in both failure and success, and, like Piaget, he takes an essentialist view in describing knowledge.

Bryant's essentialism is further illustrated by his treatment of transitive inferences:

> A child who cannot put together the information that A<B and B<C to produce the inference that A<C cannot understand even the most basic principles involved in measuring things. There will be little point . . . in teaching such a child to use a ruler, because he will have no conception that different things could be compared with each other through their common relations to it.
>
> (1974: 38–9)

This argument contains an important assumption, namely that any activity having the form 'A<B and B<C therefore A<C' entails understanding of the transitive-asymmetrical relation. But, to reverse the argument, if a child cannot make the inference as presented in a transitivity test, this does not have to mean that she cannot use a ruler. Using a ruler is an activity which concerns achievement of an end other than the making of a transitive inference. A person can measure things without representing the situation to themselves (either consciously or unconsciously) as one involving A<B and B<C. This will be because they understand the practice of using a ruler or of measuring things. According to this view, measurement is not a question of applying a formal principle to a task; rather it is a question of knowing how to act in a particular situation, of entering into the associated social practices. The corollary of this argument is that the Piagetian transitive inference task is not a task that can be understood and solved by means of abstraction of the transitivity principle from everyday measurement activities. The two situations are not linked at all except by means of formal analysis which shows that they are isomorphic in structure.

Despite his criticisms of Piaget, Bryant still takes an essentialist viewpoint which does not set him apart from Piaget in any important respects; both assume that the notions of 'transitive inference' or 'number invariance' adequately describe what someone does when they succeed in the associated tasks. Both overlook the importance of contextual understanding and the possibility that knowing how to use numbers may be a question of entering into social practices such that learning is not a question of simply putting logical competence into performance.

Gelman and Gallistel's work

Like Bryant, Gelman and Gallistel's (1978) main purpose is to present an account of children's number development which is substantially different from Piaget's. In particular, they set out to give an account which replaces the Piagetian mode of limiting the description of children's development to an account of lacunae in their thinking; they intend a more positive approach, emphasizing children's abilities as opposed to their failures. But, again like Bryant, Gelman and Gallistel fall back into an essentialist approach which prevents them from giving a positive account: essentialism entails a description of knowing in terms of the presence or absence of formal conditions prerequisite for being said to know X, and thus describes all knowing in terms of its proximity to a formal end-product. Gelman and Gallistel's final theory does not differ in substance from Piaget's, being no more than a different account of the ages at which certain steps in development are achieved.

Their work falls into two parts: first, their positive account, by which they intend to show that children are aware of the principles of counting; and second, their account of why children fail conservation tests. I shall look at each of these in turn.

The principles of counting

Watching children count, Gelman and Gallistel noticed that even when children used a totally idiosyncratic set of number tags, such as 'two, six', they would stick to this order over time and use the tags just as though they were in fact the numbers one and two: they would give six as the 'cardinal number' of the set in answer to the question 'How many?' for instance. Gelman and Gallistel take this as evidence that children as young as two and a half years can apply the how-to-count principles and that much of the development around these principles involves skill at applying them. Gelman and Gallistel can draw no support for their ideas from children who use the conventional number words and orders, since it can always be argued that they are simply reciting the products of rote learning. However, children who use idiosyncratic tags and orders provide stronger evidence, since it cannot be argued that such systems can ever have been heard by the child:

> The significant fact about these lists is that they are used in a way that is prescribed by the counting principles. It seems reasonable to conclude that the availability of the principles governs such behaviour . . . . When we first encountered such behaviours we thought them random uses of number words and the alphabet. But when we subjected them to analyses suggested by the counting principles, we discovered

that such children were telling us, in their own way, what they know about counting.

(1978: 203)

To recap: the how-to-count principles are: the one–one principle (unique tags for each and every item of the array); the stable-order principle (tags must be assigned in the same order always); and the cardinal principle (the last tag used signifies the cardinal number of the array). Gelman and Gallistel first give their reasons for attributing knowledge of these and other principles of counting to children, and then compare their work with Piaget's theory.

How to count

Gelman and Gallistel argue that analysis of the types of errors that children make illustrates their use of the one–one principle. They make two kinds of error: partitioning and coordination errors. The first kind appears when the child makes a mistake in separating off items already counted from items yet to be counted: she may double count or miss out an item, but hardly ever jumps around in the array or goes backwards and forwards. Gelman and Gallistel argue that this is because the children are guided by a partitioning rule which makes them ordinarily proceed in orderly fashion from item to item. Mistakes, they maintain, arise because of poor execution; the child is still following the one–one principle. The second type of error occurs at the beginning and end of a count: a child may start to point before she starts to recite, or vice versa, and at the end of a count may run by an item. But these are the only mistakes of this nature, for the count may proceed smoothly once it has started, and counting never runs over by more than one tag. Again, Gelman and Gallistel argue that the child realizes that tagging and partitioning should stop together, but that mistakes happen because of lack of motor skills.

With regard to the stable-order principle, Gelman and Gallistel refer to the child's tendency to use idiosyncratic lists which do not change over counts, including alphabetic lists. They say:

the conclusion seems inescapable that his behaviour is governed by a principle that specifies the use of a stably ordered list but leaves open which stably ordered list should be used . . . . The alphabet is such a list, and it is one list that young children in our culture encounter at an early age. To explain its use in counting, we postulate the availability of a stable-order principle that guides the assimilation of relevant environmental input.

(1978: 206)

Gelman and Gallistel cite as further evidence the children who used numbers, but in an unconventional order. Again, they assume that this behaviour is

100

guided by the stable-order principle. Additionally, they found that users of idiosyncratic lists were better able to follow the rule than children who used the conventional order; they argue that this is because a child will remember her own idiosyncratic list better than the conventional list which is 'imposed from outside' (207).

Gelman and Gallistel admit that their case for use of the cardinal principle is weaker, since the actual number of children acting in accordance with this was fewer, and depended on a small set size. They suggest (207) that there is a developmental lag between the acquisition and use of the one–one and stable-order principles, and the cardinal principle: successful application of the latter is dependent on skilled use of the two former principles, hence the restriction of the cardinal principle to small sets where children 'have already developed the skill to apply the one–one and stable-order principles in concert' (207–8).

The order-irrelevance principle

Correct counting, Gelman and Gallistel argue, involves not only the how-to-count principles, but also the 'order-irrelevance' principle: the understanding that the order in which items are counted does not matter. This principle relies on the prior understanding that assignment of a number word to any particular item is only temporary for the duration of the count and does not constitute naming the object. Gelman and Gallistel found that, when required to count the same array several times, children showed no tendency to assign the same number to particular items. They also asked the children to count an array of items, and then to count it again, beginning with the second or third item; alternatively, they were asked to make a particular item the number two or three, for instance. Gelman and Gallistel recorded the children's explanations for their answers, and comment that:

> Children who do well at reassigning numerlogs to a given item are quite good at explaining why they can do what we ask them to do. They invoke the principle in their verbal accounts. They have conscious access to the principle that permits them to modify their typical way of counting. If we are right in this interpretation, then what develops is insight about the principles that govern counting. Thus the hard data are strong evidence that the child comes to know what it is that he does when he counts. The child not only honours the order-irrelevance principle but likewise understands what counting is about.
>
> (1978: 218–19)

Gelman and Gallistel's protocols show that children (particularly aged 4–5 years) often referred to the fact that objects were rearranged during the tests, so that different objects could be given different number tags; a slightly different interpretation from Gelman and Gallistel's is that the children's explanations can also be seen as explicit references to convention in using number words, for instance: 'Their name can't all be number one . . . except when you're counting, they can . . . . You give them each a number but not the same number. You can change when you move them' (157); 'That's how you're supposed to do it because they're supposed to be number one each time, different times' (158).

Piaget's theory and Gelman and Gallistel's evidence

Having argued that children are able to use number at a much younger age than Piaget claims, Gelman and Gallistel must then give an account of why children cannot perform successfully in the number-conservation task, since this is taken by Piaget as the minimum criterion for possession of the number concept. Noting the importance of one–one correspondence in Piaget's analysis of the number conservation task, Gelman and Gallistel argue that knowing that two sets which are placed in one–one correspondence are equal is not a necessary prerequisite for understanding numerical equivalence. Instead, they argue, the ability to make conservation judgements is dependent on a 'later stage in the use of reasoning principles' (1978: 228). Specifically, this later stage is the stage of 'algebraic reasoning'.

Judgements of numerical equivalence

Gelman and Gallistel's argument is initially based on Gelman's (1977) 'magic' experiments which were devised in order to see if children distinguished between number-relevant transformations (addition and subtraction) and number-irrelevant transformations (substitution and spatial displacement). There were two phases to this experiment: the first established an expectation in the child of how many objects should be present in an array, and the second recorded the child's reaction to a 'magical' transformation.

The first phase entailed showing the children two plates containing different numbers of toys. The experimenter labelled one the winner, and one the loser, without referring to number. The plates were then covered and shuffled, and the children were asked to guess which was the winner, and then check to see if they were right. If a child had guessed incorrectly, and then, when the plate was uncovered, correctly identified it as the loser, she was allowed to uncover the other plate, and asked if it was the winner. The child was given immediate feedback about these identifications: if she incorrectly identified the plates when they were uncovered, the experimenter corrected her and began the game again. When the child correctly identified the winner, she was rewarded and given verbal confirmation. Thus, feedback was consequent on identification after the plates were uncovered, not on the guessing part of the game.

After at least eleven trials on phase one, phase two began. The experimenter secretly altered the winning set in terms of spatial arrangement, substitution, addition or subtraction. The child suddenly discovered that neither plate was now identical to the winner set. The experimenter asked what had happened: how many objects there had been, how many were now on the plates, if the game needed to be 'fixed', and if so, how.

Gelman recorded the children's spontaneous counting behaviour and justifications for their actions; much of her theorizing about number is based on these experiments. Two sets of results are of interest in this particular context: first, children could still judge a set to contain a certain number of items after substitution and displacement; and second, when asked to fix the game, they would often construct two sets with the same numerosity, particularly if this meant they could get two winners. Gelman and Gallistel claim that Piaget's explanation for children's failure on his conservation task – that they do not have a number concept – cannot be correct because these experiments demonstrate that children do, at least in Gelman and Gallistel's terms. They claim instead that the problem is that the child fails to use her reasoning principles, and it is this failure that must be explained.

The relationship between one–one correspondence and conservation

Discarding the possibility that failure lies in counting errors, Gelman and Gallistel suggest, in direct opposition to Piaget, that the explanation centres on the very act of using one–one correspondence to form an initial judgement of equivalence. The perception of one–one correspondence creates two problems: (i) it produces equivalence in a form that the child cannot deal with because her reasoning principles do not apply to equivalence between unspecified numerosities but only to equivalence based on having the same cardinal number; and (ii) the obvious pairing of items in the two rows distracts the child's attention from the task and from the fact that there are two rows, and thus discourages her from simply applying the counting procedure that would avoid the first problem. Gelman and Gallistel suggest a three-stage development: both of these two problems operate at the first stage, while at the second the child is unable to reason about transformations performed on a *single* unspecified numerosity, but is still unable to apply this reasoning to the effect of a transformation on the relationship between *two* unspecified numerosities. At this stage, Gelman and Gallistel argue, the child can in theory make a correct conservation judgement by following a two-step reasoning process which considers each numerosity as a single unspecified numerosity and then deduces that since the rows were initially equal, the relationship between them is unchanged. However, the fact that the rows are initially presented in pairs may distract the child from following this line of reasoning (1978: 230).

The third and final stage is achieved when judgements of numerical equivalence based on one–one correspondence are possible because the child's reasoning principles apply directly to equivalence relations between unspecified numerosities. Gelman and Gallistel call this the stage of 'algebraic reasoning'. They go on to say that 'this argument assumes that arithmetic reasoning develops from a numerical stage to an algebraic stage. Numerical reasoning deals with representations of specific numerosities. Algebraic reasoning deals with relations between unspecified numerosities' (1978: 230–1). The 'pre-algebraic child' can count two sets of four and judge them to be equal, despite displacements. But when judging two sets placed in one–one correspondence to be numerically equivalent on the basis of the correspondence and not of counting, the child, when asked to reason about the effect of displacement, 'has no appropriate input for his reasoning principles' (231). That is, Gelman and Gallistel argue, the child cannot operate without a specific numerosity as she lacks the following 'algebraic reasoning' principle: 'The equivalence between two entities, x and y, representing unspecified numerosities, is unaffected by displacing either the numerosity represented by x or the numerosity represented by y' (231). The young child's principles apply only to actual numbers, then. Gelman and Gallistel thus argue (231–2) that pre-algebraic children cannot reason about unspecified numerosities, even if they notice the one–one correspondence between the rows, because their 'reasoning principles' apply only to the effect of transformations on numerosity, not their effect on relations. Children who can conserve, therefore, are those who are able to reason algebraically in terms of numerical relations between unspecified numerosities.

Gelman and Gallistel need to present evidence for their argument by showing that young children can take note of one–one correspondence and still fail to conserve. This evidence they consider provided by Piaget's (1952) demonstration that children can place a row of items above a given one so that both rows contain the same number, but still deny equivalence when the two rows are displaced. This is weak evidence, though, since it can be interpreted in many ways: it does not constitute evidence that children are noticing one–one correspondence *per se* since they may be matching the rows according to various other criteria suggested by the context. In terms of Gelman and Gallistel's argument, it does not show that the children cannot conserve precisely because of the one–one correspondence.

Gelman and Gallistel also argue that the distracting salience of one–one pairs accounts for their observation that pre-schoolers often fail the conservation test even when the arrays are small enough for them to count: attention is drawn away from the rows themselves and their properties. They overlook the alternative explanation, though, that 'non-conservers' simply do not recognize the conservation situation in the first place; to say that they are distracted by the pairs assumes that such children understand the matter in hand but become uncontrollably side-tracked. It may be, however, that the only sense in which they are distracted is that they do not understand what is required of them, and so interpret the task as best they can, picking on what appear to be the most important features of the situation and deducing the experimenter's intention from these.

However, Gelman and Gallistel claim further support for the theory that one–one pairings have a disturbing influence from Miller and West (1976) and Markman and Seibert (1976). Miller and West's study involved varying the degree of emphasis on one–one correspondence using four different levels of emphasis ranging from high to low as follows: (i) each pair of objects was identical and unique, and was joined with thread; (ii) as for condition (i), but no thread was used; (iii) identical items in one row were paired with identical items in the other – for instance a row of bottles with a row of corks; (iv) all the items used were identical – for instance pairs of identical corks. Miller and West found that the degree of emphasis had no effect on conservation, and concluded that the equivalent performance across conditions 'raises the question of whether correspondence is even an important basis for conservation' (1976: 422–3). They also made the observation that the greater the emphasis on one–one correspondence, the less counting was observed; Gelman and Gallistel comment that this is hardly surprising since emphasis on one–one correspondence detracts from a focus on the row structure. But these results can be interpreted as counterexamples to Gelman and Gallistel's argument: instead of remaining at the same level, conservation should have increased when the emphasis on one–one correspondence decreased, particularly since counting was observed more frequently with decreased emphasis. Miller and West's evidence seems to provide only weak support, then, for Gelman and Gallistel's theory.

Markman and Seibert (1976) also argue that pre-schoolers' real numerical knowledge is masked by standard Piagetian tasks because children of this age are unable to organize the material correctly. Correct organization, they suggest, can be achieved by encouraging children to think in terms of collections and not classes. This idea is based on the assumption that collections, because of their part–whole relations ('an oak is part of a forest'), have 'greater psychological coherence' than classes, which are organized according to class-inclusion relations ('an oak is a tree'). Markman and Seibert consequently modified the standard class-inclusion task and indeed found that the class versus collection distinction had an effect on class-inclusion behaviour.

The relevance for Gelman and Gallistel's theory is that, applied to conservation, Markman and Seibert's procedure should encourage focus on the row structure rather than on the one–one pairs, and thus increase conservation. Markman (1979) pursued this application of the class versus collection distinction to conservation by using identical materials and procedure in both standard and modified conditions, but referring to the materials with either a class label (soldiers, football players, animals) or a collection label (army, football team, animal party). Again, performance improved in the collection condition, which also led to 37 per cent of children making reference to the irrelevance of the transformation as against none in the class condition. Markman concluded that performance on the standard task is not indicative of true numerical ability and that 'the children's difficulty may lie in their

inability to impose cognitively the appropriate organisation on the material'
(1979: 396–7).

Notice once again an explanation of children's behaviour which says that
they possess certain underlying principles but are unable to put them into
operation in certain situations. Markman does not recognize the possibility
that, for a child, the two situations may be substantially different in so far
as the experimenter's intention is clearer in one than in the other: possibly,
the class condition's reference to number as opposed to length is significantly
more ambiguous than the collection condition's, in which case producing the
correct answer will rely more heavily on successful negotiation of the social
cues involved. There may be contextual cues which lend support to a much
simpler explanation than Markman's 'cognitive organization' one; as I shall
argue in Chapter 8, there may be no need to postulate an essential underlying
logic common to both tasks because of their formal identity.

To summarize Gelman and Gallistel's argument, they propose that a child
will progress through the following sequence of stages before she is able to
perform correctly in the standard number-conservation task:

Stage one: Reasoning can only be applied to specific numerical values.[4]
This reasoning involves the understanding that some operations (identity
operations) do not change the numerosity of a set.

Stage two: The child applies this reasoning in the absence of a specific
value. She thus infers that the numerical value that would be obtained if the
set were counted would not change after certain operations. She must realize
that the array has a numerical value, although she may not represent that
value.

Stage three: The child recognizes that the operations in question do not
affect numerical relations. The reasoning principle then applies to relations
between numerosities rather than to individual numerosities. This is the
algebraic stage.

Gelman and Gallistel claim that the justifications given by 4- and 5-year-
olds in Markman and Seibert's studies illustrate their two earlier stages: some
children mentioned actual numbers and said that they were still the same (the
earliest stage of 'purely numerical' reasoning); while others said that the
transformation was irrelevant to number – 'you just moved them'. This
second type of justification Gelman and Gallistel believe reflects the second,
'semialgebraic' stage in which children can judge the effects of a transformation on a single unspecified numerosity but cannot apply the same reasoning
to the relation between two unspecified numerosities. At this stage, they
suggest, it is possible to come to a correct conservation judgement by reasoning that the transformed set has not changed, that the non-transformed set has
not changed, and therefore that the two sets, being originally equal, must still
be equal. Gelman and Gallistel argue (1978: 235) that this two-step process is

reflected in conservation justifications which appeal to the irrelevance of the transformation to the number of elements in the transformed row.

With respect to Piaget's theory, Gelman and Gallistel observe that

Piaget's notion of an operational concept of number and our notion of an algebraic treatment of number are closely related . . . . Our views and Piaget's part company with respect to the implications of failure to conserve. Piaget regards the failure to conserve as a sign that the child lacks a concept of number, that is, a coherent set of principles for reasoning about number . . . We, on the other hand, have argued throughout . . . that the preschooler has a coherent set of principles for reasoning about numerosity . . . . We are led instead to the conclusion that what such children lack is the ability to reason about numerical *relations*, that is, the ability to reason algebraically.

(1978: 236–7)

Gelman and Gallistel thus agree with Piaget about the reasoning which underlies successful conservation: an ability to manipulate the one–one correspondence relation. They differ from him only in the respect that they define this as 'algebraic reasoning' as opposed to 'numerical reasoning', while for Piaget it represents true number understanding, for the reasons set out in Chapters 2 and 3. The difference between the two accounts, then, lies in the fact that Gelman and Gallistel set a lesser criterion for possession of the number concept. Piaget's argument for the importance of one–one correspondence is, however, inadequately addressed by Gelman and Gallistel, and the least clear and least supported aspect of their account is their apparently self-contradictory claim that children can be distracted by one–one correspondence, while at the same time the essential component of conservation is understanding of the one–one correspondence relation.

In contrast to their earlier, more informative account, Gelman and Gallistel's treatment of the Piagetian number-conservation problem is far less useful. In attempting to explain why the Piagetian number-conservation experiment does not constitute a test for numerical reasoning, their account falls back into an implicit essentialist position: in order to solve the conservation problem, an understanding of one–one correspondence is necessary, a competence which may be blocked by certain difficulties of 'cognitive organization'. Thus there is no description or discussion of what it is that children may understand by the conservation situation and which might cause them to misinterpret the experimenter's intentions, or to treat 'classes' differently from 'collections', or to be overwhelmed by the salience of the between-row pairings, if indeed this is what happens. Gelman and Gallistel's account is a good one until they begin to deal with the area of investigation and age group that is dominated by Piaget and the majority of number-development researchers: their attempt to present a formal description of children's conservation responses represents abandonment of a positive account and a return to the 'logical lacunae' of Piaget.

This chapter has discussed the work of Piaget, Bryant, and Gelman and Gallistel from the point of view of whether, as psychological theories, they present adequate accounts of what knowing about number entails. A first step is to examine what these theorists mean by 'the number concept'. In the first section, I argued that Piaget's account is essentialist: it characterizes knowing about number in terms of certain necessary and sufficient conditions, and thus assumes that all uses of number have the same common 'essence'. Such a view can be criticized from the point of view that the many different applications of number are not united by a common principle, but in fact differ such that knowing about number is a question of entering into various social practices. The second and third sections identified essentialist approaches in both Bryant's and Gelman and Gallistel's work. I argued that neither of these theorists examine fully the implications of the research they report, either in terms of the demands made on children by the experimental situation, or in terms of the part played by context in using numbers. Like Piaget, they assume that children understand what is required of them in the experimental situation, and that they can easily apply numbers in situations of which they have little experience, provided they possess the requisite essential principles. An alternative view requires a sensitivity to how far a child can be said to have entered into the various social practices associated with using numbers, as Chapter 8 shows.

Thus both Piaget and his critics share an asocial account of knowing which, I have argued, is inadequate to explain what children know when they know about numbers. In the next chapter I will consider Doise, Mugny, and Perret-Clermont's attempts to include social factors within a Piagetian framework from the point of view of whether they succeed in deflecting the criticisms of Piaget's account of knowing raised in the last two chapters. The other major criticism of Piaget's theory is that he does not succeed in giving an adequate account of growth, as I argued in Chapters 4 and 5. In Chapter 7, I will also consider an attempt to improve on this account, again within the Piagetian framework; this is the information-processing theory put forward by Case.

7

Can a Piagetian perspective be defended?

In Chapters 4, 5, and 6 I identified two related problems in Piaget's theory. The first was the problem of how to describe and explain the growth of knowledge: despite Piaget's constructivist attempt to describe a middle way between empiricism and rationalism, his theory fails to improve on an abstractionist account. This failure, I argued, is a result of the second problem, which is that of the characterization of knowing itself: Piaget's theory assumes an essentialist conception of knowing which cannot describe what is involved in knowing number; understanding numbers is, rather, a question of entering into the social practices of their use. Furthermore, Piaget assumes that the child is a 'solitary knower' who must construct his world alone; but if, as I have argued, knowledge is intrinsically social, this account must be replaced by one in which an essentially social being acquires knowledge of number by coming to act in ways that are appropriate according to public standards of correctness.

There are, then, two major criticisms of Piaget – that he fails to explain the development of knowledge, and that he presents an asocial and essentialist account of knowing and coming to know. In this chapter I will examine further developments within the Piagetian framework and consider whether they answer these criticisms. The first section covers the work of Doise, Mugny, and Perret-Clermont, which addresses the role of socio-cognitive conflict in development. I will argue that Doise and his co-workers do not succeed in countering the epistemological objection that knowledge itself, not just its growth, must be intrinsically social. Nor, I will argue, does their socio-cognitive conflict account succeed in describing the growth of knowledge. The second section looks more closely at the question of the acquisition of knowledge, and considers the information-processing approach to Piagetian theory as represented by Case's 'neo-Piagetian' task analyses of success and failure on Piagetian tasks. I will argue that this attempt to improve on Piaget's account fails to achieve any solution to the basic problems of explaining the growth of knowledge, and that it too presents an essentialist account of knowing.

Can a Piagetian perspective be defended?

Socio-cognitive conflict and the growth of knowledge

In previous chapters I have argued that Piaget fails to present an account of knowing which expresses the fact that knowledge is intrinsically social. This epistemological criticism has to be distinguished from the weaker criticism that Piaget's theory fails to take into account the role of the child's social environment in cognitive development. Given Piaget's epistemological system, this omission is justified: Piaget has neglected the social aspects of knowledge on the grounds that knowledge is an individual construction, the product of insight into one's own actions in the physical world, and something which cannot be passed on in the social tradition, at least not before a child is ready to receive it.

Doise, Mugny, and Perret-Clermont, amongst others, have attempted to redress the balance towards individualism in Piagetian theory by reasserting the role of social interaction in cognitive development. They have resurrected Piaget's (1928, 1932, 1950) stance on social experience, an early recognition of the role of discussion which was abandoned in his later work on cognitive development. They quote, for instance, Piaget's early claims that 'social interaction is a necessary condition for the development of logic' (1976a: 80) and that 'cooperation produces reflexivity and self-awareness . . . allows the subjective and the objective to be separated . . . [and] is the source of regulation' (1976b: 114). Drawing on the early Piaget, Doise and Mugny suggest (1984: 12–13) that intelligence is more than a property of the individual: it is also a 'relational process between individuals'. They propose, then,

> a social definition of intelligence that incorporates but goes beyond a Piagetian notion of cognitive development. While Piaget describes intellectual activity as coordination, we believe that this coordination is not only individual but to an equal extent social in nature. It is in the very coordination of his actions with those of others that the individual acquires mastery of systems of coordination which are later individualized and internalized . . . [Our thesis is] that coordinations between individuals are the source of individual coordinations, and that the former precede and produce the latter.
>
> (1984: 23)

On the premise that 'not just any form of social interaction will be beneficial at any time in development' but that 'certain prerequisites are necessary to profit from social interactions' (Doise and Mugny 1984: 23), Doise, Mugny, and Perret-Clermont developed a programme of experimental study manipulating different types of social interaction. In this section I will examine their research from the point of view of whether it counters the objection that Piagetian theory fails because it ignores social factors. I will argue that, because Doise and his co-workers only introduce social factors in

110

such a way as to answer the weak criticism of Piaget's theory, they fail to address the epistemological criticism. This is because, while they introduce social factors at the point of knowledge *acquisition*, they do not do so in terms of knowing itself. Thus, Doise and his colleagues present a picture in which social interaction contributes to the evolution of individualized, autonomous knowledge. I will argue that this picture of knowledge fails.

A clear starting point at which to introduce the social environment directly into the Piagetian model is that of conflict. Recall from Chapter 4 that Piaget's theory of equilibration and consequent cognitive growth features an important role for conflict in terms of the integration and ultimate anticipation of disruption or disequilibrium. Thus the child encounters physical obstruction in the sensorimotor stage, perceptual disorder in the pre-operational stage, and incompleteness of structures in the operational stage. My criticism in Chapter 4 was that Piaget fails to give an account of how disequilibrium is experienced, and in particular how disequilibrium can be experienced by a 'solitary' knower. Doise, Mugny, and Perret-Clermont present a potential solution to this problem by introducing socio-cognitive conflict. Thus Perret-Clermont (1980: 132) claims that 'cognitive conflict created by social interaction is the locus at which the power driving intellectual development is generated', while Doise and Mugny argue more explicitly that:

> when another asserts an opposing centration to that of the child, the child is faced with a conflict not only of a cognitive but also of a social nature. This sociocognitive conflict which allows two opposing centrations to exist simultaneously cannot be as easily denied as a conflict resulting from successive and alternating individual centrations.
>
> (1984: 28)

Doise and his colleagues thus propose socio-cognitive conflict as a mechanism for growth. I will argue, however, that, just as the problem arises with Piaget's notion of intra-individual conflict that recognition of contradictions is not possible without foreknowledge of their resolution on a higher plane, so recognition of inter-individual conflict is a product of development, not a causal factor. In order to support this argument I will draw in the second part of this section on evidence produced by Russell (1982) that non-conservers in inter-individual conflict do not in fact perceive conflict to exist. I shall argue that this is because coming to experience conflict as such involves entering into social practices governed by the public criteria of correctness in the conservation-test situation. Non-conservers are unaware that the conservation situation requires them to make judgements not in terms of how things look, but in terms of volume or number. Recognition of the criteria for making correct responses cannot be a result of conflict between non-conservers.

Can a Piagetian perspective be defended?

Socio-cognitive conflict

Groups versus individuals

Doise, Mugny, and Perret-Clermont (1975) began their investigation of socio-cognitive conflict with a comparison of individual and collective performances on Piagetian concrete-operational tasks. Thus, using a spatial-coordination task designed to test awareness of perspective and decentration, Doise *et al.* found that two children working together to construct a copy of a model village performed successfully whereas age-mates working alone did not. In a second experiment Doise *et al.* looked for evidence that previous group experience on a task leads to a new understanding in the individual child. Using a variant of the standard conservation of liquid task, Doise *et al.* found significant progress amongst non-conservers who had earlier participated with conservers in a group situation; furthermore, more than half of the children introduced justification arguments which had not been used during the interaction, thus demonstrating genuine advancement. Similarly, Perret-Clermont (1980) found evidence that group performance on the same task led not only to progress in liquid conservation but could also be shown to generalize to other domains.

Doise *et al.* thus claim that interaction leads to 'elaboration of the operational structure' of the individual (1975: 381). Exactly what type of interaction is not revealed in these studies, although Doise *et al.* suggest that the most productive situations were those in which conservers have to defend their view 'in a consistent and coherent manner'; they argue further that 'these observations indicate the importance for the non-conserving child of not simply being put in the presence of equals and interacting with them but of being confronted by partners who defend a different mode of reasoning in a stable manner' (1975: 382).

Prerequisites for progress

Although the effect was not observed in the conservation studies, one of the main observations from Doise *et al.*'s (1975) spatial-coordination study was that progress seemed to occur even when neither partner was immediately able to give a correct answer. Doise and Mugny (1984) comment that 'in such groups "progress" can be linked to the existence of a conflict between individuals who possess different, though incorrect, strategies' (114). Such a link would support the view that inter-individual conflict accelerates development by presenting the child with two simultaneous opposing viewpoints. Thus later work by Doise and his colleagues concentrated on the question of what kind of interaction is beneficial in producing full cognitive restructuring and specifically on the question of conflict.

Mugny and Doise (1978) compared different interaction situations on the same spatial coordination task as that used by Doise *et al.* (1975). This time

112

they incorporated several variations on the theme of conflict. Initially, children were classified in a pre-test according to whether they used an unsuccessful strategy, an intermediate strategy, or a successful strategy; the experimental situation involved pairing children in various combinations. Mugny and Doise predicted that those combinations leading to most conflict would result in progress, and that conflict would result from the use of different strategies, as long as there was a base-line of compatibility.

With respect to the children's performance in the interaction situation itself, Mugny and Doise found that the collective performances of the unsuccessful/intermediate and intermediate/intermediate pairs were superior to those of the children working alone. Correspondingly, the post-test results showed that although unsuccessful subjects made little progress when paired with unsuccessful or successful subjects, they did make significant progress when paired with intermediate subjects. Mugny and Doise argue (1978: 189ff.) that this is because the potential conflicts of two unsuccessful children working together or of an unsuccessful child working with a successful child were diffused since, in the former case, both subjects used the same incorrect strategy, while in the latter, the successful child tended to ignore the unsuccessful child and work alone. In the unsuccessful/intermediate situation, the intermediate child is less dominant, and his doubts and hesitations lead to explicit expression of the problems involved in the task, Mugny and Doise argue, so that the initially unsuccessful child realizes what he has to do. Children who worked in the intermediate-pairs situation benefited only slightly in this study, although Doise *et al.* (1981) report progress for such pairs when the task is engineered so that the children use similar strategies, but work from different physical perspectives.

Types of interaction

Although these experiments show that children on lower performance levels may make considerable progress through interaction with children on higher levels, the exact nature of the interaction is unclear and requires further investigation if the conflict hypothesis is to be elaborated.

Accordingly, Mugny *et al.* (1975–6) examined the hypothesis that socio-cognitive conflict will result in progress, but exposure to a correct model will not – that is, progress does not result from mere imitation. In this study, an adult collaborator provided the conflict by disagreeing with the child's non-conservation responses in a similar vein to the child: for instance, if the child claimed that, of two equal-length rulers initially placed in parallel, ruler A when displaced was the longer because it extended beyond ruler B, the collaborator argued that in fact B is longer, pointing to its extension beyond ruler A in the opposite direction. In order to disallow the child from merely complying with the collaborator's view, the experimenter intervened if necessary by taking the child's side, thus prolonging the conflict and disallowing simple compliance with an adult view, since the two adults themselves disagreed. Mugny *et al.* compared this situation with one in which the collaborator in response to the child's non-conserving response gave a model correct answer with full justifications. Although both conditions produced post-test progress, Doise and Mugny (1984: 86–7) interpret the results as strongly supportive of a pure conflict model of cognitive development because a number of the children produced conservation responses despite strong advocacy by adults of non-conservation responses.

Similarly, Mugny *et al.* (1978–9) investigated social conflict by varying the intensity of conflict. As before, strong conflict was initiated by a collaborator who disagreed with the child and chose the opposite ruler as the longer, although this time the experimenter did not join the argument on the child's side if he showed signs of compliance. Instead the child was merely reminded of his initial view and asked for justification of the change. If, on the other hand, the child resisted the collaborator's viewpoint, the experimenter drew attention to the disagreement and asked if any reconciliation of views was possible. This was contrasted with a weak conflict situation in which the child and collaborator gave their (opposing) responses, after which the experimenter simply asked the child to give a final response with justification. The post-test results showed that the greatest improvement occurred in the strong conflict situation; again, Doise and Mugny (1984: 90) take this as strong support for their conflict model.

There are a number of points to note about these two studies which concern the social cues conveyed to the children by the participating adults. Whatever else they may do, these adults act on one important premise: that the conservation task is appropriately answered with one answer only. Thus, in the first study, the collaborator and experimenter prolong the conflict, indicating by this fact and by the strength of their advocacy that the situation demands one and only one answer. In the second study, the experimenter in the strong conflict situation asks the child for a resolution of the conflict; again the adult's request makes clear that one answer – not two – is appropriate; this contrasts with the weak conflict situation in which no such cue is given.

That these are important cues is shown by Russell's (1982) work, which I will discuss in detail in the next part of this section. Russell suggests (1982: 79ff.) that non-conservers are in fact tolerant of conflict in the sense that they appear undisturbed by the existence of two different answers to the same question. This can be interpreted to mean that in fact pre-operational children are unaware that the correct response in the conservation task is a judgement of sameness/difference on which all parties may agree rather than judgements of sameness/difference which appeal to how things look to different people occupying different physical positions: that is, the experimenter's intention in asking the conservation question is not to ask for judgements concerning perceptual appearance. Thus the real effect of the adults' actions in these two studies may be not so much that they present conflicting opinions, but that they convey to the child that the appropriate way to act in this situation entails giving only one answer and not two. Doise and Mugny's theory does not take into account the fact that the adult models and the forcefulness with which they are advocated represent an appeal to public standards of correctness which may bring the child to the appreciation of the force of a norm.

Social conflict and social practice

The evidence that Doise, Mugny, and Perret-Clermont present shows that, providing certain prerequisites are met, group performance is superior to individual performance and results in post-test progress. Doise and his colleagues interpret this evidence in terms of a model of socio-cognitive conflict; it is by this means that they introduce social factors into a description of cognitive development.

Notice, however, that this theory still preserves the Piagetian picture of the 'solitary knower': the end-product of the socio-cognitive conflict process is still individualized and autonomous knowledge. As I argued in Chapters 5 and 6, this characterization of knowing is inadequate to express what is involved in, say, being able to produce the correct response in conservation tasks. It ignores the fact that, in order to do so, a person must enter into the social practices which embody the public criteria of correctness in conservation judgements.

As a consequence of this, Doise and his colleagues fail to note patterns in their data which distinguish between two types of task – the spatial coordination task and the conservation task. As I will show below (pp. 116–17), these patterns are not explicable in terms of their conflict model. They can be explained, however, in terms of differences in the nature of the two tasks: while the spatial-coordination task has a clearly defined end, the ability to respond correctly in a conservation task entails that a child must enter into certain social practices and become party to public standards for the correctness of conservation judgements. Thus, when children progress on the conservation tasks, it is because a change in terms of 'tuning into public criteria' has occurred. Using Russell's (1982) evidence, I will argue that one consequence of this tuning in is the ability to see that two non-conservation judgements conflict; conversely, a child who cannot make correct conservation judgements will not in fact perceive that he is in conflict with his partner.

Thus, in the final part of this section, I will argue that Doise, Mugny, and Perret-Clermont's retention of an individualized and asocial view of knowledge means that their conflict model of growth is vulnerable to the same criticisms as Piaget's inter-individual conflict model: that is, that the perception of conflict is a *product* of learning, not its cause.

115

Can a Piagetian perspective be defended?

Does a conflict model explain the evidence?

One of Doise, Mugny, and Perret-Clermont's (1975) observations was that the greatest likelihood of progress for a non-conserver came from being paired with a conserver who clearly defended and justified their point of view. Doise and Mugny's (1984: 59ff.) interpretation of this is that the conserver's strong defence is a vehicle for the presentation of conflict which is resolved by the non-conserver by a move to conservation judgements; for Doise and Mugny, the important fact is that a conflict exists, and the stronger it is the better. However, notice that, in the spatial-coordination task this asymmetry of partners is non-productive: for instance, Mugny and Doise (1978) found that pairs of children having one child scoring very low on the pre-test and one scoring very high on the pre-test resulted in little progress for the less able child on the spatial-coordination task. Mugny and Doise suggested that this was due to dominance by the more advanced child, who prevented the less able child from joining in in any practical way and thereby learning. They also suggested that because the two children use strategies which are totally incompatible, potential conflict is diffused.[1] The conflict model succeeds, then, in explaining the results for asymmetric pairs.

However, it fails to explain the results of studies of interactions between genuine equals: that is, those in which both children are less able. Doise, Mugny, and Perret-Clermont (1975) claim progress for pairs of children of intermediate ability in spatial-coordination tasks as do Mugny and Doise (1978). By contrast, there is less evidence to suggest that two non-conservers will progress on a joint conservation task: Russell (1982) has found evidence to the contrary, while work by Doise and his colleagues suggests that progress is only possible with considerable manipulation of the situation in terms of the emphasis and prolongation of conflict. The weakness of the evidence for unmanipulated progress is acknowledged by Doise and Mugny themselves (1984: 96), while both Perret-Clermont (1980) and Zoetebier and Ginther (1978) found that groups composed entirely of non-conservers did not make progress.

The conflict model, therefore, fails to explain the evidence from studies of conservation – that is, why two non-conservers who give opposing incorrect responses should fail to benefit from the conflict existing between them. In fact, the converse of the conflict model seems to be true: there is evidence to suggest that, from the non-conserver's point of view, there is no conflict to be resolved in such interactions, while perceiving conflict between two non-conservation responses is in fact a result of being able to conserve. Thus a conserver is aware that there exist standards or norms by which the claims of non-conservers are unacceptable, because the conservation question asks for judgements on which all parties must agree; the conserver knows that it is not appropriate to answer in terms of perceptual appearance. Such a conclusion could not be arrived at by two non-conservers working alone.

Thus an alternative interpretation of Doise, Mugny, and Perret-Clermont's

116

results presents itself: children progress in interaction with a conserver because a new kind of approach is suggested by the behaviour of the other child or the adult model – the child becomes aware of the public criteria of what constitutes a correct conservation response. One aspect of this awareness will be the recognition that opposing non-conservation responses are in conflict because, while two people may have different viewpoints with respect to the appearance of the materials, the conservation question does not require a response in these terms; rather, it requires a judgement of identity despite transformations.

Does conflict occur?

One of the major assumptions made by Doise and his co-workers is that, when pairs of pre-operational children make conflicting judgements, they will indeed see these as conflicting. However, to echo the argument I put forward in Chapter 4 with respect to individuals: what guarantee do we have that these children will see that there is conflict? If a child makes un-adult-like judgements about occasions for subjectivity versus objectivity, then it may be the case that: (i) 'conflict' is not really conflictual; and (ii) we have a poten- tial explanation for pre-operational performance in Piagetian tests.

Thus, if two pre-operational children take different (and incorrect) views on the relative contents of two jars, Doise, Mugny, and Perret-Clermont's idea is that the children will make claims and counterclaims, arguing like adults who are sure they are right, in favour of their particular point of view. Eventually, this scenario suggests, the children will recognize that neither of them is right and that a compromise along the lines of conservation is the only single answer. Doise and his colleagues assume that children in such a situation will perceive conflict, but this entails that, as Hamlyn says, they will be party to the public standards of what counts as true and objective and will therefore know that, in this situation, judgements made from two subjective viewpoints are incorrect and that one correct, 'objective' answer is required. If they are unaware of this requirement, however, they cannot perceive their judgements as in conflict. As Russell notes:

> The children would not engage in a dialectic at all: the conversation over their glasses would have more in common (in logicality if not style) with that of two wine tasters considering whether 'tart' or 'piquant' is the appropriate adjective.
>
> (1982: 79–80)

Russell provides some empirical support for this argument: in a length- conservation situation arranged so that two pre-operational children will take opposite perspectives, Russell recorded the children's conversations and assessed these in terms of the attitudes expressed and the amount of conflic- tual behaviour in evidence (claims, counterclaims, etc.). The data suggested that, in fact, non-conservers 'did not appreciate the fact of cognitive conflict' (82). The production of conservation answers was very low and overall Russell reports little evidence for perspective coordination of the kind suggested by Doise, Mugny, and Perret-Clermont.

What in fact happened was that the children gave as an answer that proposed by the most socially dominant child of the pair (assessed by their tendency to 'take over' in a joint picture-drawing task and a story-completion task). Close inspection of the interactions suggested that this choice of answer was not so much due to strong advocacy on the part of the more dominant child as to fast compliance by the other child. The children's justifications often made reference to situational factors (e.g. 'my brother is good at these and he told me . . .' (Russell 1982: 83)) and, even more strongly supportive of the argument that non-conservers judge correctness in terms of how things look and are unaware that this is not what the conservation task asks for, sometimes made reference to perspective as support ('it *looks* bigger', 'that one's bigger to you and that one's bigger to me' (Russell 1982: 83)).

Non-conservers do not, apparently, realize that there is a conflict because they do not operate in terms of there being one right answer. In contrast, conserver–non-conserver pairs, in which the 'winner' was typically the conserver, produced interactions in which the conservers gave more opposing judgements, and defended their position more vigorously, giving justifications in terms of rules of reversibility and so on. The relationship between 'winners' and social dominance witnessed in the non-conservation pairs did not hold for the conserver–non-conserver pairs: the conserver won whether or not he was dominant in the partnership. This, Russell observes, is because 'he regards his answer as *true* and his partner's as being *false*' (1982: 82). By contrast, the non-conserver does not have 'a notion of "correctness" versus "incorrectness" that is sufficiently like our own. Perhaps he is giving something that we would call an opinion when we would like him to make an attempt at "the answer"' (75).

Recognizing public criteria

The fact that non-conservers do not recognize conflict supports the argument that, in fact, coming to conserve involves appreciating the force of a norm in terms of 'doing things the way *they* do' and so coming to know the 'right' answer. As Russell suggests, development is 'the progressive socialization of the child's judgements, their tuning in with those of the adult by means of the appropriate public criteria' (1982: 78). By contrast, Doise, Mugny, and Perret-Clermont take the development of understanding and knowledge to be a question of a social path to *individualized* knowledge: 'Autonomous expression of cognitive competences is only possible as a result of development based on social interdependence . . . . From this interdependence arises a cognitive collaboration which gives rise to cognitive competences allowing the subsequent autonomisation of development' (Doise and Mugny 1984: 157).

But Russell's studies demonstrate what Doise and Mugny's studies and analysis fail to show: that knowledge is intrinsically social, and that knowing, as well as coming to know, hinges on social experience. Social interaction is not merely a conflictual catalyst for what is ultimately individualized, 'autonomous' knowledge, to use Doise and Mugny's phrase; rather it is the vehicle by which a person enters and participates in the social practices that define the public criteria of what is to count as knowledge.

For Doise, Mugny, and Perret-Clermont, conflict leads to development, with an end-product of individualized knowledge in which social experience plays an evolutionary part only. Russell's work supports the contrary view that conflict in the sense of claims and counter-claims is a symptom of development that has already occurred: that is, development towards participation in practices which define the public criteria of what is to count as right and wrong.

Information-processing theory and Piagetian theory: Case's 'neo-Piagetian' analysis

One of the major problems in Piaget's theory is his failure to describe and explain the development of knowledge in detail. As a consequence, a number of information-processing theorists – for instance Case (1978a, 1978b, 1982, 1985), Siegler (1978, 1986), and Klahr and Wallace (1976) – have set themselves the task of directly addressing the gaps in Piaget's theory, which Piaget himself later recognizes (see Piaget 1978b, for instance) as potentially removed by means of an approach which lays emphasis on procedural rather than conceptual knowledge; information processing entails such an approach.

In information-processing theory 'thinking *is* information-processing' (Siegler 1986: 63). A theory aims to produce explicit 'task analyses' – step-by-step break-downs of what the successful performance of particular tasks such as class inclusion,[2] say, require: 'which processes are performed, in what order and for how much time . . . what information is represented in what form and for what period of time' (Siegler 1986: 63); it also aims to give similar step-by-step accounts of children's failure given such tasks. Additionally, such a theory seeks to present an account of children's cognitive capabilities and limits at particular ages and to explain the change mechanisms involved in the development of adult memory and reasoning processes and the cognitive limits imposed on the speed of such development in children. Thus, for Case, whose 'neo-Piagetian' analysis is the subject of this section, cognitive development is a matter of increasing automatization and the consequent by-passing of limits in working-memory capacity. I shall argue, however, that his theory is unsuccessful in achieving its aim: the failure of the information-processing approach to progress beyond an abstractionist model, and its tendency to characterize children's thinking in

119

terms of 'logical lacunae' in the adult model is clearly shown in Case's analysis.

Case's 'neo-Piagetian' analysis

Case's account is clearly related to Piaget's. Specifically, he retains Piaget's conception of the major stages of development each of which is characterized by a qualitatively different underlying operation. He also maintains, with Piaget, that the process of development is a question of equilibration. He criticizes Piaget, however, on the grounds that his theory does not easily become a theory of instruction (1978a). In particular, he argues, Piaget's theory does not provide the means for: (i) the analysis of what is involved in performing particular tasks successfully; (ii) the analysis of what failure to perform a task successfully involves; or (iii) the construction and design of teaching methods which will bring a child from failure to success. Case's neo-Piagetian approach, which is based on a theory originally put forward by Pascual-Leone (1970), locates as the source of these deficits the fact that Piaget's theory is presented in structural rather than functional terms. Accordingly, the theory is a version of Piaget's account of development but is influenced by artificial intelligence models (Simon 1962; Newell and Simon 1972). Thus Case's approach differs from Piaget's in that it conceptualizes knowledge in terms of sets of operations which govern the processing of information in particular tasks. It also differs in its introduction of the notion of working-memory capacity as the major developmental factor; this is the key innovation in Case's theory.

Originally designated as 'M-power' or 'M-space' by Pascual-Leone (1970) and Case (1978a), working-memory capacity is defined as 'the maximum number of independent schemes that can be attended to at any moment in the absence of direct support from the perceptual field' (1978a: 195). Thus the notion of working-memory capacity works on the assumption that the activation of a mental scheme requires a certain amount of 'mental energy' (188). Specifically, working memory is constituted by a constant amount of processing space which is occupied by: (i) information about what is to be done in a particular task and how it is to be accomplished; and (ii) the problem information itself. According to the neo-Piagetian analysis, all tasks can be characterized in terms of the number of elements that must be manipulated and coordinated if a problem is to be solved.[3] Particularly important is the question of how many schemes must be attended to at the same time: if a child's working-memory capacity is exceeded by a particular task's memory demands, then he will not be able to perform that task.

Case maintains that each person has a central working memory which is used as the space for both storing information and working on it and that the basic capacity of this working memory does not change. (Case differs here from Pascual-Leone (1970), who maintains that actual memory capacity increases with age from an M-power of one at age 3–4 to five at age 11–12.) But within each stage there is an apparent increase in capacity which is due to a decrease in the working-memory space required to execute particular tasks. This functional increase in capacity enables development to occur in the form of the acquisition of more complex executive schemes. The reason for this is that there is an increase in the automaticity of basic operations; when this occurs, carrying out the operations requires less working-memory capacity than before, hence releasing space for the acquisition and use of more complex schemes. This process will be recapitulated in the next stage, as in Piaget's theory: transition is consequent on a sufficient degree of automaticity being acquired in the previous stage to enable assembly in working memory of the new underlying operation from components of the previous stage.

The notion of working-memory capacity also enables Case to make a further innovation in terms of the value of instruction in cognitive development. In Piagetian theory, the role of instruction is minimal, but Case's introduction of the notions of working-memory capacity and automatization into the Piagetian framework entails a greater potential for instruction in development. Specifically, instruction can enable a child to acquire an executive scheme at a younger age than normal by reducing the memory demands of the task. Thus Case maintains that there is a difference between situations which merely present an opportunity to learn and situations which involve direct instruction. Presented with situations in which there is an opportunity to learn, Case argues, a child will develop more sophisticated strategies only if working memory has the capacity to cope with additional information; on the other hand, explicit instruction which draws the child's attention to new cues and strategies will have the effect of reducing demands on working memory by means of automatization or of 'chunking' together items which are normally attended to separately. Hence, Case argues, explicit instruction can be effective, providing it is correctly tailored to the child's working-memory capacity.

Task analysis and the processes of development

Case (1978a, 1978b, 1982, 1985) provides task analyses for a number of Piagetian tasks, specifically seriation (1978b: 46–7), classification (1985: 207–8), counting (1982: 162–3), and also conservation of liquid (1978a: 178–86). I shall concentrate here on the latter analysis as the most detailed model of knowing and coming to know.

Task analysis, Case (1978a: 205) suggests, can be undertaken by means of introspective analysis of the operations and sub-operations executed in order to reach a particular goal. Occasionally, he owns, such introspection may be difficult 'because the answer to the problem question may appear to leap into one's mind without any mental steps at all' (206): such immediate insight tends to occur with the conservation task, for instance. The solution, Case maintains, is to 'explore the problem from a purely rational point of view and to *set out a series of steps that would logically be necessary in order to prove the answer that appears to be obvious*' (206, my emphasis). Case goes on to say that it is precisely this method that Pascual-Leone (1972) used for his original task analysis of liquid conservation. Thus the successful liquid conservation strategy is as follows (1978a: 181):

1. Recall the original relative quantity of the two beakers. Store.
2. Recall the nature of the transformation.
3. If the transformation was simply pouring, state that the final relative quantity is the same as the original one.

Note, however, an objection to this method: a rational reconstruction as to why a conservation judgement is correct does not necessarily correspond to the actual process by which someone comes to be able to perform the task correctly. Case presents here what may be called the opposite of the genetic fallacy: he takes the logical steps by which something can be said to be true and assumes that these are the steps by which someone comes to understand it as true. This criticism arises in a different form with respect to Case's conflict mechanism of change as I will show below.

Also necessary to the investigation of any Piagetian task is an analysis of strategies that fail. In Piaget's system, failure on particular tasks is reduced to logical lacunae, as I showed in Chapter 3; this is no less true of Case's system in which this tendency becomes very clear:

> Because the analysis of an incorrect strategy is more difficult than the analysis of a correct strategy, it is often helpful to be able to convert the former problem into the latter one . . . . One useful technique is to determine a question for which the child's response would have been correct and then to cycle through [the analysis] as though this were really the goal of the task . . . . An alternative technique that is equivalent to the first one is to *assume that the subject does not have access to some crucial piece of information that he appears to have ignored*.
>
> (1978a: 208, my emphasis)

Thus Case explicitly states that a child's failure to correctly perform in, say, conservation tests is a result of missing information: otherwise all else remains the same.

How the child comes to apprehend this missing information, under normal circumstances, is by a process of recognition and resolution of conflict. Thus the incorrect strategy immediately preceding the successful conservation strategy above is one in which the child takes account of the difference in the height of the two beakers and also the difference in width and makes a judgement of which beaker has more according to whether a difference is 'large' or not. However, use of this latter strategy can lead to conflict: the child may find that if he uses height in order to make judgements he gets a different answer from that which he arrives at if he uses width:

1. Note that Beaker B has taller column of liquid than Beaker A.
2. Conclude that Beaker B contains more water by this criterion. Store.
3. Note that Beaker B is narrower than Beaker A.
4. Conclude that Beaker B contains less water by this criterion. Store.
5. Note that the two stored conclusions are in conflict.

(1978a: 182)

Case suggests that 'the experience of this conflict might lead him to actively search for some new dimension and to work out a new strategy for taking it into account' (181). Thus he claims that the experience of such conflict 'activates a heuristic executive scheme that directs a search of the problem situation, the problem given, *and any stored information of relevance*' (187–8, my emphasis):

1. Remember that the two original beakers (A and B) started out with the same amount of water in them.
2. Remember that the water in Beaker B was transformed into the tall thin beaker (B') by pouring, without any water being added or subtracted.
3. *Drawing on past knowledge* of such pouring acts, conclude that the amount of water in B was unchanged in the course of its transfer to B'.
4. Putting together the fact that B' is still equal to B, and that A was equal to B to begin with, conclude that A is equal to B'.

(1978a: 182, my emphasis)

The problem with Case's explanation of change is that it presupposes not only that the child has recognized the value of bidimensional scanning as opposed to unidimensional scanning (making judgements solely on the basis of the height of liquid), but also that he will recognize the conflict between judgements based on height and judgements based on width. As the discussion of Doise, Mugny, and Perret-Clermont's and Russell's work in the previous section suggests, this will not necessarily occur, either in terms of intra-individual or of inter-individual conflict.

On a more detailed level, the question arises as to how the heuristic executive strategy is to 'direct a search' for relevant information and clues without already knowing what it is looking for, and here we see the explicit use of an abstraction model in Case's theory: a striking feature of the strategy above is that it sounds less like a description of the discovery of conservation than an argument for it such as an adult might give, and indeed, as I have already shown, this is exactly what the origins of the strategy are. Thus the strategy assumes an abstractionist and question-begging picture of coming to know: abstraction of new knowledge from past experience (see my emphasis, p. 123) relies on that past experience being appropriately categorized and thus presupposes the new concept itself. As with Piaget's system, the recognition and resolution of conflict tends to presuppose the more advanced knowledge state which is supposed to be the *product* of a recognition of conflict, not its *cause*. Coming to respond correctly in the conservation task involves more than the mere apprehension of additional information: it involves the recognition that the correct response is one which distinguishes between considerations of appearance and considerations of number, mass, and volume. Similar criticisms apply in Case's analysis of the missing-addend problem, as I will now show.

Neo-Piagetian instruction in the missing-addend problem

Case's neo-Piagetian theory of instruction involves, as I noted at the beginning of this section, three aspects: a task analysis of the task to be learnt, an assessment of the child's current level of functioning on the task, and the actual instructional design itself. Instruction, in Case's system, involves demonstrating to the child the inefficiency of his existing strategy and then helping him to construct a new one. At the same time the familiarity of the situation and also the saliency of the relevant cues must be maximized in order to minimize the memory demands of the new strategy and keep this within the limits of the child's current working-memory capacity. The final move is to consolidate the new strategy by means of practice so that it becomes automatic and makes lesser memory demands; the strategy can now be extended and complications introduced. Case (1978a) provides an illustration and empirical study of this procedure. He focuses on the difficulty of the missing-addend problem for young children, and describes a teaching programme devised to enable the children to perform the problem given their limited working-memory capacity.

Missing addend problems require that one fills in the gap in equations like the following:

$$4 + ? = 7$$

It is well documented (see Chapter 8) that children first attending school experience difficulty with this problem type but that, as a rule, they produce answers which are not random but consistent with one of two strategies: either they give four or seven as the answer; or they produce the answer eleven. Case observes that children who have had very little training in arithmetic tend to give the first type of answer, while those who have had more training tend to give the second type.

To begin with the initial task analysis, Case suggests that the task demands are as follows:

1. Read the symbols from left to right.
2. Notice that the quantity to be found is one of the two addends.
3. Decide that the addend that is known must therefore be subtracted from the total (which is also known).
4. Note the value of the given addend. Store.
5. Note the value of the total. Store.
6. Subtract the addend from the total.

(19778a: 214)

According to Case the crux step is number (3) which presupposes that the child realizes the reversibility of addition and subtraction and, additionally, demands an M-power of from three to five (214). He argues that if it were possible to devise a strategy which did not require this reversibility of addition and subtraction, then it should be possible to teach it to younger children who have a lower M-power. He suggests that the following is such a strategy:

1. *Read the problem from left to right, looking for the operation sign (+).*
2. *Look at the symbols on either side of the addition sign.*
3. *Note that one of the numbers to be added is given, but the other is not (indicated by the presence of a box).*
4. *Find the equal sign and note the number on the other side of it.* Store this number (7).
5. Begin with the number after the given addend and count in increments of one (5, 6, etc.). For each number counted, advance a token (e.g. a finger).
6. When the stored total is reached (7), stop.
7. Count the number of tokens advanced and enter this number in the box (3).
8. (Optional) Reread the symbols from left to right, checking the verbal statement against one's knowledge of number facts to be sure that it is correct (e.g., Does 4 plus 3 equal 7?).
9. If the statement does agree, stop. If it does not re-enter the task at step 4.

(1978a: 215, my emphasis)

There is, however, a way of distinguishing between the two strategies which Case overlooks. Case claims that the important and difficult step of the first strategy is step three, because it presupposes an understanding of the reversibility of addition and subtraction which is omitted in the second strategy. This may, however, be complicating the issue: successful solution of the problem may not rely on such Piagetian reversibility, but, rather, on the understanding which underlies step two, 'notice that the quantity to be found is one of the two addends'. Considerable knowledge of the conventions of arithmetic is necessary for this step: it requires isolation of an unknown, an awareness of the problem syntax, and an ability to interpret it, and also the recognition that this particular problem is not like those we meet most commonly, that is, $4 + 7 = ?$. It is likely to be the case that children in the initial stages of learning arithmetic tend to follow general procedures which cause them not to recognize this particular pitfall, and they will lack the familiarity with arithmetical notation and syntax rules which allows correct problem analysis and retrieval of known number facts.

It is precisely these aspects of arithmetical procedure which are highlighted in the second strategy (see my emphasis). Steps one to four are totally concerned with the details of the problem syntax: isolation of the operation to be carried out, isolation of the unknown, isolation of the givens, attendance to the placing of the equal sign and its relation to the other symbols in the written problem. It is this spelling out of the problem syntax and its interpretation which constitutes the important difference between the two strategies and in fact conveys the real understanding required for successful solution of the missing addend problem – an understanding which simply appears in shorthand in the first strategy. Case's view, however, seems to be that the second strategy does not require the problem solver to really know what they are doing: he assumes that it simply involves a less demanding strategy which by-passes the need for understanding the reversibility of addition and subtraction. Arguably, though, this conceptual knowledge has no part in successful solution of the missing-addend problem whereas familiarity with the symbols, syntax, and goals of arithmetic, together with known number facts, does. And, significantly, it is these very aspects of arithmetic which Case (unwittingly) sets out to teach in his instruction programme.

According to this analysis, when children cannot solve the sum '$4 + ? = 7$', it is not because they do not understand the reversibility of addition and subtraction, but because they do not grasp the syntax of the problem or pay attention to its symbols, although they do have access to the number facts (at least in the case where they respond '11'). Thus Case rightly suggests that the incorrect strategy which produces the answer '11' proceeds as follows:

1. Look at the first symbol. Store.
2. Count out that many tokens (e.g., fingers).
3. Look at the second symbol. Store.

126

4. Count out that many tokens.
5. Count out the total number of tokens advanced. Store.
6. Write this number in the box.
7. (Optional.) Check to see if the two smallest numbers add up to the biggest number (Does $7 + 4 = 11$?).
8. If it does, stop. If it does not, re-enter the problem at step 1.

(1978a: 215–16)

Consequent on this identification of an incorrect strategy, Case (1978a) designed and tested an instruction programme which, he maintains, accesses the child's abilities at the right developmental level and M-space demands. As I have already argued, however, the value of the programme lies not in its supposed teaching of a less demanding strategy, but, rather, in the fact that it gives explicit coaching in the symbolism, syntax, and interpretation of the missing-addend problem. The major steps of the instruction programme are illustrated in Figure 7.1.

The purpose of the first step (7.1a) is the creation of a procedure by which the child can evaluate his current strategy. The aim is to enable him to check for himself 'whether or not he has attained his objective' (1978a: 217). But Case merely assumes that the child knows what that objective is, and thus overlooks the effect of the repetitive definition and redefinition of the equal ($=$) and addition ($+$) signs as a contribution to the child's knowledge of the purpose of arithmetic and its procedures.

The next step (7.1b) is intended to demonstrate to the child the inadequacy of his initial strategy. It is unclear whether it does in fact do this or not, since the faces situation is sufficiently unlike the standard missing-addend problem to allow the child to overlook the parallels (in fact it would be useless to attempt to use the incorrect strategy since this would necessitate slotting a full face into the space for half of a face. This visually obvious impossibility would be unlikely, it seems, to lead the child to attempt the incorrect strategy in the first place.) Looking at the second picture in step (b), however, note the emphasis it places on the unknown and its role and function in the arithmetic puzzle. Thus step (b) reinforces and furthers the coaching begun in step (a).

The third step, which is not illustrated, is to explain to the child why his initial strategy is incorrect. Such an explanation may be something like the following: 'Remember, this ($+$) says put these two together so they will be the same as that one. When you put these two together do you get the same as that?' (1978a: 218). Case observes that 'this draws children's attention to the relevant cue that they have ignored ($+$)' (218); more accurately, though, it draws their attention to its usage in the wider context of the arithmetic problem, thus again referring explicitly to arithmetic convention.

The fourth step in Case's programme (also unillustrated) is to teach the child the correct strategy. Here there is a curious methodological departure

Figure 7.1 A neo-Piagetian approach to the missing-addend problem

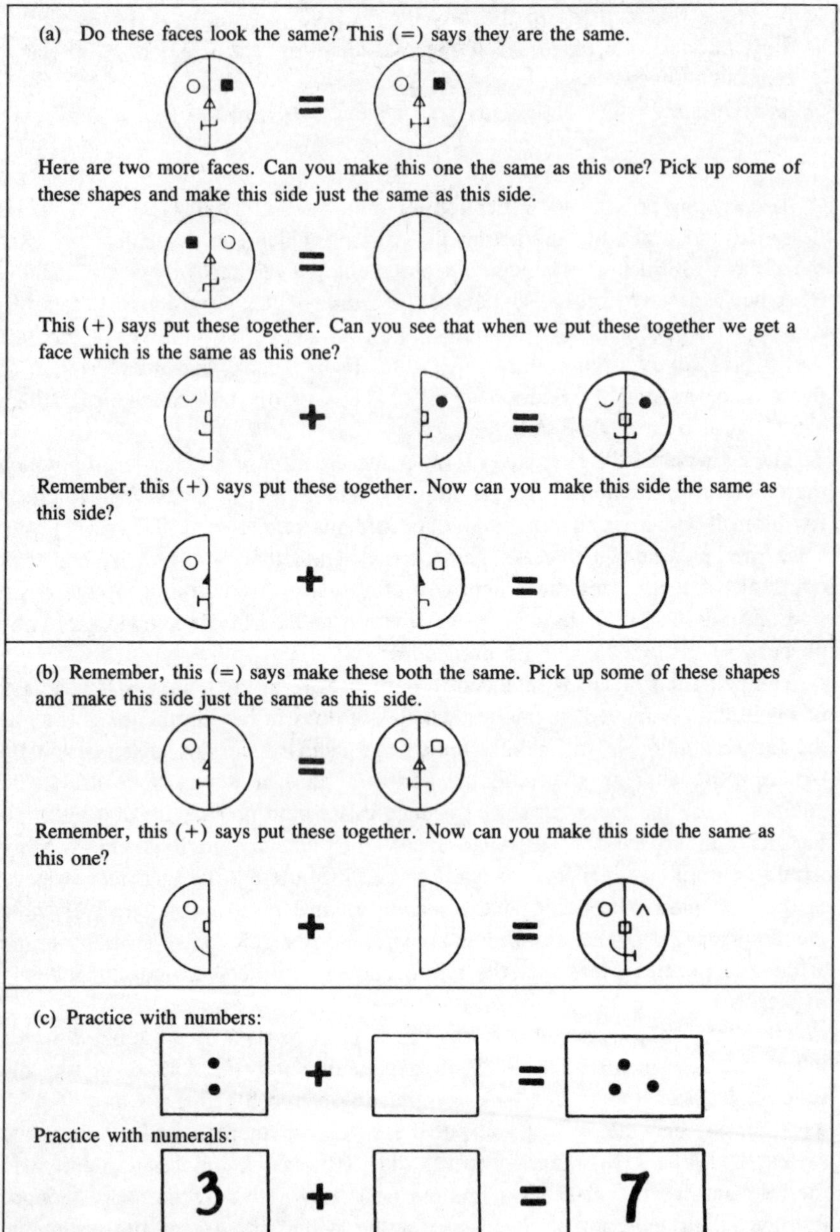

(a) Do these faces look the same? This (=) says they are the same.

Here are two more faces. Can you make this one the same as this one? Pick up some of these shapes and make this side just the same as this side.

This (+) says put these together. Can you see that when we put these together we get a face which is the same as this one?

Remember, this (+) says put these together. Now can you make this side the same as this side?

(b) Remember, this (=) says make these both the same. Pick up some of these shapes and make this side just the same as this side.

Remember, this (+) says put these together. Now can you make this side the same as this one?

(c) Practice with numbers:

Practice with numerals:

Source: adapted from Case (1978a: 216–18).

128

in Case's study: he does not in fact teach the correct strategy (i.e., counting-on from the given addend to reach the sum) for the following reasons:

> Since the face paradigm deals with very concrete materials, this is unlikely to be necessary at this stage. In principle, however, one could draw the child's attention to the way the complete pattern can be broken down into parts, and each part can be checked to make sure that it is the same as one on the other side of the equal sign.
>
> (1978a: 218)

Once again, then, Case in fact concentrates on the symbolism and the syntax. He does not teach the counting-on strategy *per se* and, arguably, does not need to because once the child has understood the procedures, purpose, and symbolism of arithmetic, the major obstacle to solving the missing-addend problem – its interpretation – is overcome. By not teaching the strategy itself, Case obscures what evidence there may be for his own analysis: it becomes unclear what, exactly, the success of the programme is due to, if not to the fact that Case has made explicit some of the conventions of arithmetic throughout the instruction programme.

Finally, the fifth stage (7.1c) involves 'consolidation and transfer'. At this point, Case says, the instruction programme reintroduces complications in the form of numbers of spots and finally numerals, produced one at a time in order to keep the memory demand of the task to a minimum. Notice again, however, the emphasis on notation: 'As each complication is reintroduced, one can drop back and retrace one's way through the original series of steps, reminding the child first of the meaning of the equal sign and then of the meaning of the plus sign' (218).

Case (1978a) reports that a curriculum based on this teaching programme was in fact successful: 80 per cent of children receiving neo-Piagetian training were successful in the post-test, compared to 10 per cent of children in a control group. As I have shown, however, the reasons why this was so are debatable; the suggestion that what in fact happened was that the children benefited from the intensive coaching with respect to the conventions of arithmetic is supported by work which I shall consider in Chapters 8 and 9 and which further illustrates that understanding how to do arithmetic is not a question of gaining new conceptual knowledge but of entering into particular social practices.[4]

In this chapter I have considered work within the Piagetian framework from the point of view of whether it answers the two main criticisms of Piaget's theory: that he wrongly excludes social factors from his characterization of knowing and coming to know, and furthermore that he fails to give an adequate account of the growth of knowledge. Thus in the first section I considered the work of Doise and his colleagues, who attempt to bring a consideration of the child's social environment into the picture of cognitive development by means of a theory of socio-cognitive conflict. I argued, however, that while this theory takes social factors into account in the acquisition of knowledge, it nevertheless fails because it retains a conception of knowledge itself as individualized and autonomous. Specifically, I argued, their emphasis on conflict fails for the reason that the perception of conflict is something which is a product of knowledge acquisition, not one of its processes.

This problem is also evident in Case's neo-Piagetian approach which I considered in the second section. Here I argued that Case's theory does not succeed in improving on Piaget's account of the growth of knowledge: in continuing to describe children's understanding in terms of adult knowledge with omissions, his analysis also shows how a conflict model of growth cannot work without presupposing knowledge of the resolution of conflict on a higher plane. Relatedly, Case's analysis in terms of memory demands highlights his failure to acknowledge the essentially social nature of knowledge and the role of context in understanding. Thus, two further points are raised by the work discussed in this chapter: first, the part played by social and linguistic context in psychological experiments is illustrated both in Case's and in Russell's work; and second, Case's study of the missing-addend problem demonstrates that learning how to do arithmetic is a question of being able to use its conventions and hence enter into its social practices. These points form the subject of Chapter 8.

8

Knowing how and when to use numbers

In Chapter 6 I questioned whether the Piagetian notions of number conservation, transitive-asymmetrical relations, and class inclusion necessarily represent what children know, either when they are doing the associated tasks, or when they use number. It can be argued that knowing about number does not take the form that Piaget attributes to it; instead it involves entering into the social practices involved in using numbers. Thus the possession of knowledge about number does not necessarily have to be shown by demonstration of the possession of certain logical ideas which conceptual analysis leads Piaget to propose as the necessary and sufficient conditions for having that knowledge. Instead, knowing about number is a matter of knowing how to act appropriately in situations in which number is used, that is to say, understanding and sharing in what other people understand and expect in those situations, and therefore knowing how and when to use numbers.

The first section of this chapter considers research by Donaldson and her co-workers on the linguistic, non-linguistic, and social demands of Piagetian tasks and the argument that these create difficulties for children with the result that their true logical and numerical competence is masked. The second section widens the discussion to include the use of numbers in arithmetic, while the third concentrates on work by Steffe *et al.* (1982, 1983), which illustrates the importance of understanding the context of 'doing arithmetic' and its associated social practices.

Linguistic, non-linguistic and social contexts, and psychological experiments

The criticisms of Piaget made by Bryant and Gelman and Gallistel and reviewed in Chapter 6 suggested that failure to perform appropriately in Piagetian tasks is due to the fact that they demand skills other than the logical or numerical competence that Piaget is interested in. Thus Bryant argues that limitations in memory capacity account for failure to perform transitive inference and that the use of inappropriate strategies for judging number leads to the failure to respond correctly in the conservation task. Gelman and Gallistel on the other hand argue that conservation tasks are 'at a minimum a task for logical capacity, the control of attention, correct semantic and estimation skills' (Gelman 1969). Similarly, Trabasso *et al.* argue that correct performance in the class-inclusion task requires 'perceptual, semantic, referential, quantification, comparison and decision processes' (1978: 151) over and above competence in logical operations.

In this section I will consider criticisms made in a similar vein that Piaget has underestimated children's cognitive competence largely because his tasks make certain linguistic, non-linguistic, and social demands which cause the child to perform badly. This criticism suggests that these extraneous demands of concrete-operational tasks lead to false negative results: an underestimation of the child's logical competence. A number of studies have therefore been undertaken with the aim of demonstrating this competence in the 'pre-operational' child. There is a countercriticism and defence of Piagetian methodology however which argues that these studies in fact produce false *positive* results: the studies in question make it possible for children to respond appropriately without recourse to logical operations at all. I will argue, however, that both this argument and the initial argument against which it is set are misconceived for the reason that the presence or absence of an underlying logical competence in children's thinking is not the issue: the question as to whether or not they may be said to have entered into particular social practices is.

Are Piaget's results false negatives?

Linguistic factors

Early criticisms that the results of Piagetian tasks were due largely to linguistic difficulties were offset by the observation that children were easily able to use and understand potentially problematic words such as 'more', 'less', 'same', and 'different' in other tasks. However, later work established that children's comprehension of such words may indeed vary from one context to another.

Thus Donaldson and Wales (1970) studied the use of the word 'same' and suggested that in fact there are a number of possible interpretations of this word. Apart from its 'standard' reference to the identity of an object over time, 'same' can also be applied when: (i) two or more objects are alike with respect to all observable attributes, such as form and colour; (ii) two or more objects are alike with respect to at least one such attribute, but different with respect to at least one other; and (iii) two objects are alike in some respect that cannot be directly observed such as function (239–40).

Donaldson and Wales studied 3½- to 4½-year-old children's understanding of the words 'same' and 'different' using two groups of objects. In group A any two objects taken at random from the group would be either the same with respect to both form and colour, or different with respect to both form and colour. In group B any two objects would be either: (i) the same with respect to colour/form but different with respect to form/colour; or (ii) different in both respects. For each group, the children were shown a stimulus object: their task was either to pick out an object which was 'the same in some way', or to pick out an object which was 'different in some way'. Donaldson and Wales found that the children showed a tendency towards picking objects which were the same. Given the first group, they picked an object that was the same in all ways, regardless of the instruction. Given the second group, they picked an object which was both the same and different in response to both instructions, although they could have picked an object which was different in both respects in response to the second instruction. Donaldson and Wales concluded that 'different in some way' for these children implied 'a denial of object identity along with the presence of some sort of similarity' (1970: 250).

Similarly, Walkerdine (Walkerdine and Sinha 1978; Sinha and Walkerdine 1978) compared children's performance on the conservation test with that on a test of compensation. In the latter, the child was shown a standard glass containing orange juice for herself and also a tall, narrow glass; this was for the experimenter's drink. The experimenter poured squash from a jug into this glass, telling the child to shout 'stop' when they both had 'exactly the same to drink'. The results showed that the majority of 3½-year-olds succeeded in the compensation test, but at age 4 failed the test and continued to do so at 4½ years old. By comparison, all age groups failed a standard conservation test. Walkerdine and Sinha (1978: 169) suggest that the 3½-year-olds achieved compensation because, for them, 'same' refers only to identity, whereas by the age of four their understanding of the word 'same' also includes attributional similarity, with the result that they favour a response in terms of levels of liquid since the situation suggests that this is the salient attribute.

Donaldson and Balfour (1968) also investigated children's understanding of 'more' and 'less'. Again, they observed differences in the application of these terms and devised a study accordingly. In study one, two cardboard apple trees were used as stimuli. Each had six hooks which held cardboard apples. The task was to 'make it so that there are more/less apples' on one of the trees, given a certain number on the other. Study two, six months later, used just one tree and the task was to 'make it so there are less apples on the tree'. In the first study the children did not appear to differentiate 'more' and 'less', responding always as though they had been asked to make it 'more'. Moreover, when they were asked to add or subtract apples they tended to add apples. Similarly, in the second study, they tended to add, not subtract, apples, although Donaldson and Balfour report that there was, apparently, some confusion as if the children knew they were doing something wrong, but were unable to correct this. Donaldson and Balfour concluded that the children were treating 'more' and 'less' as synonymous judgements of quantity and therefore as judgements of 'more'.

Clearly, such findings suggest that linguistic factors must play a part in children's performance on Piagetian tasks and that their understanding of certain words will vary between contexts; similarly, researchers have also investigated how the surrounding non-linguistic context may affect the interpretation of the test question.

Non-linguistic factors

The argument that non-linguistic factors make a difference to the child's understanding of Piagetian tasks is best illustrated by McGarrigle *et al.*'s (1978) study of class inclusion. According to Piaget, the logical rules of class inclusion are not self-evident to children under the age of six or seven who, while they concentrate on the whole class (flowers), are unable to simultaneously consider its subclasses (red and white flowers, for example). Simultaneous consideration of whole and parts is only possible, Piaget says, when thinking becomes operational and decentred. Thus when children make the standard error, they compare subclass with subclass. However, McGarrigle *et al.* suggest that this is not because they cannot do otherwise, but because this is what they think they are supposed to do: the perceptual contrast between the subclasses (red versus white flowers) encourages children to assume that the task requires a comparison of these highly distinctive groups. If this is the case, they argue, performance should improve if this assumption is discouraged and the task requirements clarified by emphasizing the total class and de-emphasizing the subclasses.

In order to do so, McGarrigle *et al.* used four model cows, three black and one white, which were laid on their sides so that they were 'asleep', thus introducing another perceptual feature of posture, as well as that of colour. Asked the question, 'Are there more black cows or more sleeping cows?', 48 per cent of 6-year-olds answered correctly compared to 25 per cent who were asked the standard question about the same array. McGarrigle *et al.* note, however, that when the cows were 'standing up' and the children were asked, 'Are there more black cows or more standing-up cows?', there was no significant improvement. They therefore concluded that changes in the wording or in the form of the array do not by themselves affect performance, but together they do.

Despite the manipulation of linguistic and perceptual variables in the 'sleeping cows' experiment, a number of the children were still unable to answer the modified question correctly. Thus McGarrigle *et al.* examined the question of whether the children's performance was specifically to do with the class/subclass comparisons necessary for class inclusion, or whether it

Knowing how and when to use numbers

Figure 8.1 Display used by McGarrigle, Grieve, and Hughes (1978) to Study the Effect of Perceptual Contrast in Class-inclusion Tasks

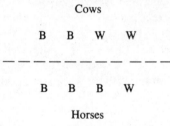

Cows

B   B   W   W

— — — — — — — — — —

B   B   B   W

Horses

B = black, W = white

was due to the misleading effect of the perceptual contrast in the task materials alone. They therefore looked at performance on similar question forms in situations which did not involve class inclusion. In one such situation the children were presented with an arrangement of black and white model horses and cows as in Figure 8.1. When asked the question, 'Are there more cows or more black horses?', only 14 per cent responded correctly. There is no question here of class inclusion, so Piaget's explanation does not fit these results. However, McGarrigle *et al.* recorded that, given this particular question, the children were apparently comparing black horses with black cows. In this particular task, then, an explanation of the children's answers cannot lie in their ability or inability to grasp the logic of class inclusion since the arrays do not involve class/subclass relations. Because of the similarity between the answers children gave in these tasks and those they give in genuine class-inclusion tasks, McGarrigle *et al.* argue that poor performance in class-inclusion tasks is not due to cognitive confusions but to the salience of the perceptual contrasts together with the question form; thus Donaldson comments that:

> the questions the children were answering were frequently not the questions the experimenter had asked. The children's interpretations did not correspond to the experimenter's intention; nor could they be regarded as normal, given the rules of the language. The children did not know what the experimenter meant; and one is tempted to say they did not strictly appear to know what the language meant. Or, if that seems too strong, one must at least say that something other than the rules of the language was shaping their interpretations – something perhaps like an expectation about the question that would be asked, an expectation that could be influenced by the nature of the experimental material.
>
> (1978: 49)

Thus Donaldson argues that children in the class-inclusion situation are influenced by the material situation and the expectations this gives rise to – that certain questions regarding the comparison of subclasses will be asked. The strength of such expectation overrides or influences their interpretation of the actual test-question language.

Similarly, Donaldson and Lloyd (1974) looked at judgements of whether statements made by a toy panda were true or false. The materials were four garages, and a set of three or five toy cars. Children were asked to judge statements such as

(A) 'All the cars are in the garages.'
(B) 'All the garages have cars in them.'

If there are three cars, then A is true and B is false. If there are five cars then A is false and B is true. Donaldson and Lloyd found some strange results: some of the children said that both A and B were false when there were three cars, and that both A and B were true when there were five cars. Such answers tally with attention to the statement, 'All the garages are full', and the children's other remarks were in accordance with this interpretation. Donaldson argues that:

> Watching the children and listening to them, one had the powerful impression that the empty garage was somehow *salient* for them, and that they interpreted everything they heard in ways affected by this salience. So we have to take into account that what the child expects to hear is liable to be influenced not only by things which give him clues about the speaker's intentions but also by more impersonal features of the situation he is considering.
>
> (1978: 67)

Thus, Donaldson argues, the child is wholly concerned with the idea of parking cars in garages, and judges the panda's statements in terms of the cars that ought to be present, rather than in terms of those that are.

Social factors

Just as Donaldson and her co-workers argue that non-linguistic factors may lead a child to have certain expectations of what a task demands, they also suggest that social or behavioural factors may influence these expectations and similarly mislead a child. This possibility is most evident in an examination of the conservation task.

According to Piaget, failure in conservation tasks is a result of failure to decentre, and of irreversibility of thought. However, Donaldson proposes an alternative explanation:

Suppose that the child is not concerned to weigh especially what the words of the question mean in isolation. Suppose he is rather interpreting the whole situation: what the experimenter says, what he does, what he may reasonably be thought to intend. Now recall that at stage two the experimenter draws attention to an action whereby he changes the array . . . . 'Watch this', he says. Is it not then reasonable that the child should think this change will be relevant to what follows – to the next question which will be asked?

(1978: 63)

Working from such a supposition, McGarrigle and Donaldson (1974) devised an experiment to judge how much weight children give to the fact that the experimenter deliberately moves the materials and comments on this move, thus suggesting that the change is important. In this experiment, the moving of materials in the second stage of the conservation test was made to appear accidental – it was not the experimenter's doing, but that of a 'naughty teddy'. Thus when the experimenter repeated the initial question on equivalence, her motives were not suspect as in the standard test. Many more 4- to 6-year-olds responded correctly under these conditions, a result replicated by Nielson and Dockrell (1982) and corroborated by Gelman (1969) who trained children to ignore the change as irrelevant; this training was also generalized in some cases to other conservation tasks.

Light *et al.* (1979) showed that the same effect could be generated when the conservation transformation was made to appear incidental rather than accidental. Thus child and experimenter participated in a game in which each was to begin with the same amount of pasta shells in similar jars. Having obtained the child's agreement that the starting amounts were equal, the experimenter 'noticed' that one of the jars was chipped, thus necessitating a swop with a different shape jar. Seventy per cent of the children agreed that they had still the same amount of shells, compared to 5 per cent in the standard condition.

A slightly different explanation of children's failure on the standard task was given by Rose and Blank (1974), who suggested that the experimenter's repetition of the question might lead children to assume that their first answer was wrong, much as happens in classroom situations. They thus decided to omit the first, pre-transformation, question, since repetition might be taken as a cue to alter the first judgement. Six-year-olds did indeed make fewer errors on this task, and also performed better a week later on the standard task.[1]

The experiments reviewed here were designed to show that, given the right circumstances, children can demonstrate the logical competence which Piaget argues they do not have. Thus they may not use individual words as adults do, they may be led astray by perceptual variables, and they may be led by unintended social cues to misconstrue the experimenter's intention in asking the question in the first place. Thus, Piaget's critics argue, his standard tasks produce false negative results. What, then, is involved in responding correctly to the standard tasks?

Donaldson argues that:

> One way to describe the difference between child and adult would then be to say that it lies in the amount of weight that is given to *sheer linguistic form*. The question seems to be whether the meaning of the language carries enough weight to over-ride the meaning of the situation. Does the language have priority? Can it over-ride the reasonable expectation?
>
> (1978: 63–4)

Donaldson's point is, therefore, that children and adults differ in that: (i) the child's knowledge of language and her confidence in this knowledge is less, so that she leans more heavily on non-linguistic cues; and (ii) the child is not easily able to attend to language in its own right. She argues further that when young children seem to understand what is said to them, they are relying on cues other than language; their skill with language alone is not very good. She maintains that children are unable to show their logical competence because they are overwhelmed by the experimental materials and situation to a degree that they cannot attend to the language of the test question and separate it from the non-linguistic aspects of the situation.

Thus Donaldson's analysis is that children are wholly concerned with what she calls the 'human sense' interpretation of the materials, and their understanding of the questions asked is coloured by this. The adult, on the other hand, is able to 'disembed thought from context' to use Donaldson's phrase, and so attend to the 'pure linguistic form'. What children have to learn to do is to 'conceptualize language':

> It is clear that being aware of language as a distinct system is relevant to the business of separating what is *said* from what is done or from what is somehow salient in a situation – and hence to dealing successfully with Piagetian tests like conservation or class inclusion and with many other reasoning tasks.
>
> (1978: 93)

In the third part of this section, however, I will argue that these ideas represent a misinterpretation of the results reviewed here.

Do Piaget's critics produce false positive results?

Smith (1982b, 1986) claims that there is a flaw in the argument that Piagetian tasks produce false negative results because the studies which support it attribute possession of a particular concept to a child on the basis of lesser criteria than Piaget himself accepts. To illustrate this view, recall Gelman and Gallistel's assertion that children as young as two-and-a-half are able to count. As I argued in Chapter 6, their account in fact differs very little from Piaget's, except that Gelman and Gallistel accept as a criterion for under-standing number the ability to act appropriately according to the how-to-count principles. Understanding one–one correspondence is, they say, a question of algebraic reasoning. There is, then, no dispute over the facts, but over the labelling of the facts. This is Smith's point too: he maintains that

> children who have a counting ability are not thereby to be credited with an understanding of *number* . . . . What matters is not simply whether a child has the ability to master number at an extensional level but also how a number is comprehended at an intensional level . . . . If the conditions extracted from the philosophical analysis are not met, there is no evidence for the attribution of a capacity.
>
> (1986: 204)

Thus Smith takes the strongly Piagetian (and clearly essentialist) view that having the number concept entails the understanding of the logic of class, correspondence, and asymmetrical relation (1986: 197–200). He also accepts the genetic epistemological aim of complementing philosophical analysis with genetic data and he defends Piaget's methodology on this basis. He argues that Piaget's method of 'critical exploration' is invulnerable to the criticism that it ignores variables such as language, context and so on because 'it is not experimental in the psychologist's sense . . . it is baldly stated by Piaget that he has "no interest at all" in the investigation of such variables' (198). Rather, Smith points out, 'the method is observational in that Piaget's primary aim is to observe whether or not a child's understanding is such that the conditions elaborated in the philosophical analysis are met' (198–9). That this is indeed the case is shown in Chapter 2.

Smith (1982b) presents the same argument in detail with respect to class inclusion. He points out Piaget's stipulation that logical competence in the understanding of class-inclusion relationships entails not only that the child should judge as true the observation that, say, there are more flowers than red flowers, but also that she should recognize the necessary truth of this statement – that it must be so. Studies which purport to demonstrate logical competence in younger children must, therefore, demonstrate the same appreciation of necessity. Furthermore, recall from Chapter 4 that Piaget's account of the child's understanding of class relationships entails that such an

139

appreciation means that the members of classes are to be identified not only by their positive qualities but also by their negative qualities.

Thus, as I showed in Chapter 4 (pp. 60–1), for Piaget, class inclusion entails the realization of the negative statement that for every class A there is a complementary class non-A and therefore that, given a superset B with subsets A and A', then, while A = B − A', it is also the case that A' = B − A. Thus, to use Smith's example (1982b: 269–70), if B = nine flowers while A = seven daisies and A' = two roses, it is also possible to express the negative properties of each subset such that A = flowers that are not roses (B − A') and A' = flowers that are not daisies (B − A). As I showed in Chapter 4 it is the apprehension of such negative properties and hence the necessity of class-inclusion relations which distinguishes the concrete-operational child from the pre-operational child in Piaget's system.

Smith proceeds to apply these Piagetian conditions for the understanding of class inclusion to those studies which claim that younger children do in fact have such logical competence. His first point is that the understanding of necessity can only be accessed via the original class-inclusion question, 'Are there more members of A or more members of B?': coming to the correct conclusion that B has more than A entails recognition of the members of A' in terms of both their positive properties (roses) and their negative properties (flowers that are not daisies); it also entails recognition that B has more members than A', which involves recognizing A in terms of positive properties (daisies) and negative properties (flowers that are not roses). It is this recognition of both negative and positive properties, Smith claims, that is not demanded by experiments such as Markman and Seibert's (see Chapter 6) and McGarrigle *et al.*'s (1978), with the result that children answer correctly at a younger age. He argues (1982b: 271) that the sleeping cows experiment, for instance, allows children to relate the classes on the basis of positive properties only:

	Class A	=	three *black* cows
	Class A'	=	one *white* cow
	Class B	=	all four *cows*
But also	Class C	=	four *sleeping* cows

Thus the test question, 'Are there more black cows or more sleeping cows?' is answered by calculating C − A, hence avoiding the necessity of understanding that some members of class B are not members of Class A and that A' = B − A.

Conversely, Smith argues, McGarrigle *et al.*'s horses and cows experiment, in which the materials are arranged so as *not* to refer to an inclusion relationship, nevertheless requires an understanding of the necessity of class inclusion. Recall that there are four cows, two black and two white, and four horses, three black and one white. The children are asked, 'Are there more cows or more black horses?' Smith argues that, to answer correctly, it is necessary to understand that the class of cows is made up of the two subsets and therefore that although the comparison asked for is not a class-inclusion one, a successful response depends on the child's ability to count the class of cows, and this in turn depends on an understanding of the class-inclusion relationship. Thus, Smith maintains, the poor results on this particular question are due to the absence of an appreciation of the necessity of class-inclusion relations.

Thus Smith's argument is that the facts are not in dispute, but the description of them is. Gelman and Gallistel, McGarrigle *et al.* and others simply differ from the Piagetian view in terms of what they take to be the minimum criteria for having the number concept, understanding class inclusion, understanding conservation, and so on. For the Piagetians, the criteria are higher, leading to higher age ranges for possession of the concepts in question, while for the others they are lower, with correspondingly lower age groups.

Linguistic and social contexts of psychological experiments

In the preceding parts of this section I have described two opposing viewpoints with respect to children's logical competence. The studies reviewed in the first part claimed to demonstrate greater logical competence in so-called pre-operational children than Piaget gives them credit for. However Smith's argument in the second part showed that many such studies simply use a lower criterion than Piaget for the attribution of particular concepts to children. Hence, Smith argues, they confuse low-level competence with high-level (complete) competence which includes an understanding of logical necessity.

In this final part of the section I shall argue that, while Smith's analysis of the modified tasks and their results might be correct, neither Smith nor the researchers referred to at the beginning of this section (pp. 132-8) in fact address the real issue of what is involved in being able to perform Piaget's tasks and their modified versions correctly. They all assume an essentialist view of knowledge which maintains that one can state the necessary and sufficient conditions for being said to know something: they merely differ in terms of what those necessary and sufficient conditions are claimed to be and/or how they can be demonstrated. Thus Donaldson, with Smith and Piaget, assumes that the ability to perform tasks like class inclusion accesses an underlying logical competence; her major concern is to show that pre-school children have this competence. I will argue, however, that Donaldson gives a wrong interpretation of the results presented on pages 132-8. This misinterpretation is founded on an essentialist view of knowledge and on a failure to see cognitive tasks as governed by particular social

141

practices which children may not be party to, rather than a question of the possession of logical competence.

As we have already seen, Donaldson's interpretation of the studies reviewed earlier in this section is that children fail on the standard tasks because they are led by contextual cues of various material and social kinds to misunderstand the experimenter's intention with the result that they misinterpret the test question or even ignore it altogether. The reason why they do this, Donaldson maintains, is that their understanding of certain words is not complete and, furthermore, that their confidence in their understanding of language is correspondingly lacking. The ability to answer the standard question correctly is, Donaldson claims, an ability to attend to the 'sheer linguistic form' of the question and to 'disembed thought from context'. Modified tasks produce success because they provide contextual support which enables the child to understand exactly what the experimenter wants her to do. To summarize, then, Donaldson assumes that understanding the experimenter's intentions is a question of attendance to the pure linguistic form of the question and no more, something which older children and adults have the linguistic ability (and confidence) to do. Contextual cues are something to be overcome.

Thus Donaldson claims that attendance to social cues in, say, the standard conservation task, is an example of immature behaviour: the child does not attend to the words in isolation but, rather, takes her cues from the experimenter's behaviour – what she does, and what she appears to intend. But this kind of pragmatic inference-making is a feature of human communication in general (Grice 1975); it is not the prerogative of children. Thus while a child may interpret the post-transformation repetition of the test question in the conservation task as indicating that a change of answer is desirable, discounting this interpretation of what the experimenter intends is not a question of disembedding language from context. On the contrary, it is a question of choosing another, equally valid, interpretation that is also based on contextual cues. These cues are ones which derive from a situation of testing and which indicate that the post-transformation question is something of a pseudo-question: it asks for information that the experimenter/tester already has and wants to make sure that their interviewee has.

The two interpretations of the question are, in a sense, equally valid; what makes one right is the public agreement regarding its appropriateness in this situation, a situation which is associated with certain social practices. It is being party to those social practices that enables someone to give the correct response to the conservation question; this correctness is not and cannot be the result of attending to 'pure linguistic form' because the meaning of the words in the experimenter's question is constituted by the behaviour which surrounds their use: they do not have a meaning in isolation.

So the child does attend to the language of the test question; she merely interprets it differently and, it turns out, inappropriately, according to public standards of correctness. The effect of the modified tasks, then, is not to enable the child to put into performance an otherwise obscured logical competence (Donaldson's interpretation), nor is it to enable her to give an appearance of competence which she does not in fact have by means of guessing, response bias, and perceptual salience (Smith's interpretation). On the contrary, the modified tasks cue the child in the direction of the right interpretation of the test question, and remove cues which lead to the wrong interpretation. They indicate what is and what is not appropriate behaviour, given that question in that situation. Development, then, becomes a question of entering into social practices, of knowing how to act appropriately in certain situations; the psychological test situation is one which we might reasonably expect a child to be unfamiliar with and to act, therefore, as a truly naive subject.

This analysis receives some support, paradoxically, from a study which in fact reports a failure to generalize McGarrigle and Donaldson's (1974) conservation results. Miller (1982), following on from the success of the naughty teddy's accidental transformation and Light *et al.*'s (1979) incidental transformation, looked at children's reactions to 'ecologically natural' transformations – boats floating apart, cars running downhill, children and crickets moving closer together or further apart, sweets being scooped up and put in a bag, and so on. Sometimes these transformations were obviously caused by the experimenter, and sometimes not. Miller compared these modified tasks with standard versions using the same materials but found no difference in performance between standard and modified tasks. However, he also carried out another study in which he contrasted replications of McGarrigle and Donaldson's and Light *et al.*'s experiments with two new versions of these; this study revealed some important contrasts between situations in which the child expects to use numbers, and those in which she does not.

Miller's study was set in the context of a Halloween game in which both the experimenter and child had a toy pumpkin into which they were to throw sweets, starting off with the same amount each. Miller's new version of the accidental transformation occurred when the experimenter knocked over her sweets with the result that they spread across the table, allowing the introduction of the post-transformation question as a check on the equality of the number of sweets owned by experimenter and child. The new incidental transformation involved the experimenter asking the child for help in setting up the game for the next child. This time two rows of sweets were lined up ready, but then the experimenter 'remembered' that, in fact, the game should start off with the child's sweets in a bag. She therefore gathered the sweets into a transparent bag, allowing repetition of the conservation question. These versions of the transformation were contrasted with a replication of the procedure used by Light *et al.*, and a partial replication of the 'naughty teddy', presented this time as a toy witch who similarly escapes from a box and 'spoils the game'.

This time the results did show superior performance on modified versus standard tasks with the exception of the witch version of McGarrigle and Donaldson's accidental transformation. So again, Miller found that using a modified conservation transformation does not necessarily lead to improved performance. (Similarly, Miller (1976) found that the results of a non-verbal test of conservation were not unambiguous.) His study did, however, isolate an important difference between the successful and the unsuccessful modifications in both studies; as Miller points out, the majority of the modified tasks provided only loose analogies with real-life situations: while it is usual to see other children move about, it is not usual to have one's attention drawn to that movement, or to be asked about the number of children before and after the change. Similarly, Miller argues, the witch is introduced as someone who might 'spoil the game', and 'indeed the only apparent purpose of her actions, and of the experimenter's subsequent question, was the possibility that spreading the candies might in fact change the number and thereby ruin the game' (1982: 229). Thus the nature of the transformation is irrelevant; what matters is whether or not the task appears to deliberately focus on numbers and on a transformation that is not normally connected with numbers.

To put these observations another way, it is only when the test question is posed in a situation in which the use of numbers is familiar to the child that she knows how to act appropriately. Only when it makes sense to talk about numbers and to repeat the test question will the child understand why the experimenter does so; Miller's version of the accidental transformation makes sense in this way: the context is one of a game in which both players must have the same number of sweets, and the accident that occurs appears genuine. In contrast, situations such as the naughty teddy and the witch do not appear at all natural: they are still engineered situations in which the post-transformation question is asked merely for the sake of asking. So repetition of the test question may be taken to indicate that something has changed because of the transformation even if it is performed by some agency other than the experimenter because the situation is no more than a test situation, the rules of which the child does not know.[2]

That children need to understand the social practices of the test situation is also demonstrated in a closer look at class inclusion. What is special about the class-inclusion question is that a correct response to it entails understanding why someone should ask such a question, and recognizing that, in a sense, it is not a genuine request for information: again, the test situation involves a kind of pseudo-question the answer to which the asker knows. Adults and older children understand such questions and know how to deal with them, but a younger child might not, and will interpret the question as best she can – as a genuine request for information. So it is misleading to say, as Donaldson does, that the child is influenced by the nature of the materials and, again, that this is immature behaviour: what else, in fact, is the child to do given that she assumes that a genuine question is being asked? Similarly, if we are looking at a field of cows and my friend asks me, 'Are there more Friesians or more cows?' I am likely to interpret this as a genuine request for information on the assumption that: (i) my friend is not trying to trick me; and (ii) she does not know the names of other types of cow. Again, the question, 'Are there more children or more people?' at a party may indicate a slip of the tongue but it is not interpreted as an exercise in logic: in an IQ test it is.

The use of 'more' to indicate a comparison question has a 'normal' usage in that we generally use it to compare subclasses, not a superclass with a subclass. Accepting these latter two as comparable and answering the class-inclusion question accordingly entails being familiar with the kind of situation in which such a question would be asked: the meaning of 'more' is constituted by the behaviour which surrounds its use.[3] Its misinterpretation in the class-inclusion situation does not mean that the child ignores the question language, nor does it mean that she lacks an understanding of the word 'more'; rather, it means that she does not recognize the particular use of comparisons which this situation requires and thus answers as best she can: indeed, as Hughes and Grieve (1980) showed, children will attend to, and try to make sense of, the bizarrest and most meaningless of questions, for instance, 'Is red heavier than yellow?'. Understanding how to respond appropriately to the class-inclusion question entails, then, understanding why someone should ask such a question.

The same is true of the cars and garages situation studied by Donaldson and Lloyd (1974). Donaldson (1978) suggests that the children's answers are based on 'human sense' about the use of garages: they should have cars in them. A mature response, on the other hand, would entail attending to the pure linguistic form and answering accordingly, Donaldson assumes. But, as in the case of conservation and class-inclusion questions, answering appropriately is not consequent on the apprehension of the underlying logic, but on choosing the appropriate interpretation of the words from a choice of equally valid semantic interpretations. A person achieves this by sharing judgements and expectations concerning what is being talked about: giving the correct answer in the cars and garages experiment entails discounting the interpretation that we are talking about all the cars that ought to be there, and instead talking about the cars that are in fact there. Again this is done not by attendance to sheer linguistic form but by recognition of the experimenter's intentions to test one's knowledge by means of questions which must be answered literally and without reference to practicalities (cf. Scribner's (1977) evidence that such an approach to problem solving is limited to literate societies).

That recognition of the situation as a test situation is necessary in order to elicit the correct response to a cars and garages type question is illustrated in a study by Freeman, Sinha, and Stedmon (1982), who observed that when adults respond to a question which entails comprehension of quantifiers, they respond to what they perceive as their interlocutor's purpose. For instance, when they are laying the table for dinner, someone is unlikely to respond to the question, 'Have you put all the knives and forks on the table?' in terms of all the knives and forks that are kept in the drawer. Freeman *et al.* therefore questioned adults about two pictures of domestic scenes. In one, four saucepans were paired with four saucepan lids plus one other lid lying next to the cooker among other kitchen items. In the other, four saucers were pictured, three of them having a cup on. The questions were asked quite casually, in a non-testing type of situation: the subjects were told that their help was needed to check the materials for a psychological experiment on children's language comprehension. The twenty subjects were asked, 'Are all the lids on the saucepans?' or 'Are all the cups on the saucers?'. Sixteen replied 'yes' to the first question and seventeen replied 'no' to the second, thus replicating children's 'errors' in the cars and garages experiment.

On the assumption that the adults concerned understood the rules of semantics, Freeman *et al.* conclude that their responses were dependent on the perceived purposes and intentions of the speaker: they answered in terms of 'all the cups that ought to be present', and in terms of a distinction between functional and non-functional saucepan lids. Freeman *et al.* therefore suggest that the conventional rules for interpreting quantified expressions rest on the assumption that if a speaker asks about the presence of 'all the Xs', they are implicitly asking the hearer to check that no X is missing, and, further, they are implying that indeed some X or Xs might actually be missing. Effectively the situation biases the hearer towards giving a negative reply. It also suggests that it is not relevant to mention extra referents such as the spare saucepan lid: answering a question about 'all the Xs' is therefore likely to entail ignoring irrelevant Xs and searching for cues which indicate a missing X such as an empty garage or a cupless saucer. This analysis leads once again to the suggestion that, in the cars and garages experiment, children are construing the experimental situation in such a way that a genuine information-seeking purpose can be attributed to that particular experiment; effectively, they misinterpret the pseudo-question.

Recognizing the experimenter's real intention entails recognizing the situation for what it is and answering according to its particular social practices. Thus Donaldson's view that older children and adults succeed by disembedding thought from context is mistaken. On the contrary, children's failure to solve class-inclusion and conservation questions is not a result of the blocking of competence by context, but instead is the result of insufficient understanding of the context and the way that certain words are used in it. Knowing how to use number and solving Piagetian tasks is actually constituted by the understanding of a situation and its associated social practices.

The context of doing arithmetic

In the preceding sections I have argued that knowing about numbers is a question of knowing how and when to use them, and is not the product of an individual construction dependent on the expressive power of the child's underlying logic; it is instead the result of an initiation into social practices which involve numbers. The concept of number does not have a unitary meaning such that knowledge of numbers can be attributed to someone on the basis of their meeting certain necessary and sufficient conditions which are specified beforehand. By this account, learning about numbers is not a question of learning certain essential characteristics of numbers, nor is number something that can only be apprehended by an individual human mind at a specific stage of logical development. In short, it can be argued that number use is not necessarily abstractable, self-evident, or inevitable, but has to be learnt from others by means of entering into social practices and learning to do things as others do in particular situations. Learning something in respect of how and when to act entails an understanding of the relevant contexts, and this is the case however abstract is our use of number; thus even the most advanced mathematician operates within a context, the understanding of which is intrinsic to the knowledge involved.

Evidence such as that presented by Tizard and Hughes (1984), Hughes (1981, 1983, 1986), and Gelman and Gallistel (1978) has shown that pre-school children are able to use numbers in counting, labelling, telling the time, playing cards, chanting number rhymes, and informal addition, subtraction, and division. In so far as they act within a known context, children know what they are doing and why: Tizard and Hughes (1984) record 4-year-olds who can recognize number symbols, know which numbers are bigger and smaller than others in a whist-type card game, count items on a shopping list, subtract within the context of a number song, add up the number of cakes required for a certain number of people, and share out cakes among a number of people. It follows that, similarly, when children do arithmetic, they have to know what they are doing and why: there is no necessary reason why they should abstract the understanding of arithmetic from their everyday play with objects and use of numbers, for these may be only formally related; there is not necessarily any abstractable form underlying everyday number use which will eventually come to the surface in the shape of mathematical rules and mathematical understanding.

It seems likely that, for a child to see those rules and to understand the connection between spontaneous 'real life' actions and the questions which adults ask (whether they refer to abstract or concrete manipulations), she has to understand the context of 'doing arithmetic'. This is more than saying that the connection between abstract and concrete representations has to be made vivid; before this can happen, the features of doing arithmetic at all, both abstract and concrete, must be understood. These may include the methods of representation and calculation used, the need for accuracy, the purpose of generating abstract rules, and the purpose of arithmetic itself as a method for finding out particular kinds of things. Arithmetic calculations can have everyday applications, but often are far removed from these, and so demand a particular kind of attention to detail and an understanding that arithmetic problems can be an end in themselves.

This is an aspect of doing arithmetic which is often overlooked by those who study it. For instance, Resnick and Ford (1981) and Resnick (1982) concern themselves with the debate over whether teaching arithmetic should centre on 'procedural or conceptual issues' and the link between abstract and concrete representations, while overlooking the fact that none of these makes sense without an understanding of the context of arithmetic. Similarly, discussion of 'discovery learning', through which the concepts of arithmetic are supposed to be deduced, frequently overlooks the fact that this method of teaching requires that the child does not discover just anything, but, in this case, the rules of arithmetic. As Dearden says, 'the upshot of a lot of bustling activity might be confusion, muddle, uncritical acceptance of first ideas, or failure to have any ideas at all, as well as possibly having the result of making a discovery' (1967: 142).

Informal activity to do with numbers and manipulation of objects outside of school learning is generally spontaneous and undertaken with an instrumental purpose in mind. This activity is to be contrasted with the formal number use which takes place in the context of doing arithmetic; here the purpose is to apply formal rules to number in the context of studying numbers and number problems for their own sake. This view does not however entail that it makes sense to equate concrete manipulation with informal number use, or abstract ideas with formal number use: the contrast between formal arithmetic and informal number use cuts across that between concrete and abstract representations. Thus it is possible to have concrete representations in an arithmetic problem, but these are still *representations*, and it is necessary to see them as such, as created for a particular pedagogic purpose. In contrast, abstract ideas can enter into informal use: children might spontaneously refer to number-relevant hypothetical events in the future or past, ask what time it will be when . . . , and so on. Thus formal arithmetic can be both abstract and concrete, and so can informal number use.

The idea that children might learn arithmetic just by being in the world assumes: (i) that ordinary activities have underlying them formal structures which are easily abstractable; (ii) that these structures can be and are abstracted with no more ado by children in learning arithmetic; and (iii) that arithmetic exists as a unique, independent domain waiting to be discovered. The problem with this notion is that it assumes that arithmetic as an object of study is a subject which involves a simple observation of the facts. But, as I will show in the next section, doing arithmetic involves particular methods which follow a particular social tradition; participation in the activity of doing arithmetic requires that someone should be familiar with its context. As Dearden points out, the idea of arithmetic as simple observation and abstraction makes no sense, since 'the concepts and truths of mathematics are not even empirical, and hence, can [not] plausibly be represented as wide open to the curious gaze of tireless young investigators' (1967: 144). And further:

> When a teacher presents a child with some apparatus or materials . . .
> he typically has in mind some one particular conception of what he
> presents in this way. But then the incredible assumption seems to be
> made that the teacher's conception of the situation somehow confers a
> special uniqueness on it such that the children must also quite inevitably
> conceive of it in this way too, even though they may not even possess
> the concepts involved . . . . A special potency is thought to inhere in
> teaching apparatus . . . if children play with it or manipulate it, signifi-
> cant experiences must be had, and important concepts must be
> abstracted.
>
> (1967: 145–6)

Simple presentation of concrete materials cannot, in point of logic, make an idea clear to someone who does not already have that idea:

> We may well ask why it is that although children have played with
> blocks and bricks for years, it is only now, when they are provided in
> schools, that important mathematical concept-abstractings are supposed
> to accompany play with them.
>
> (1967: 147)

Such a view assumes a conception of the child-as-scientist, which attributes to the child so much of the adult way of conceiving of the world that it begs the question of how a child becomes initiated into adult conceptions.

The idea that children could abstract formal arithmetic from their ordinary actions and play without some sort of guidance is, by this account, incoherent. Such guidance would involve more or less explicit explanation or demonstration of the purpose and methods of arithmetic: what it is for, how and why it is used, and how it relates to ordinary activity. Even when using concrete materials, the fact that, in arithmetic, objects are used to *represent* things, separates doing arithmetic from ordinary activity, even when they are the same kind of objects. Thus the concrete enactment of 'Mary has eight buttons. She gave six to her mother. How many does she now have?', for instance, requires a child to know that she has to pick eight buttons from the pile in front of her, take away six of these, and count the remainder. In doing this, the child must have an idea of using materials as representative of objects in hypothetical problems devised for the purpose of explaining or testing arithmetic skills. She must also understand that doing the sum involves setting up the initial set of eight and ignoring the fact that there are more left over in the pile because this is an arithmetic class and these are not really Mary's buttons, and Mary does not really exist.

Understanding context is important even when a child has some arithmetical knowledge already. Thus, as Dearden points out, the arrangement of Cuisenaire rods to represent the sum $2 + 4 = 6$, according to their normal use and according to the teacher, relies on particular conventions of representation, so that the alternative possibilities of $2 + 4 + 6 = 12$, or $\frac{1}{3} + \frac{2}{3} = 1$, or $40 + 80 = 120$ are ignored. Thus, even within mathematical interpretations, there are many equally valid possibilities in any situation. So, as Dearden says:

> 'Mathematics is all around us' the advocates of this sort of discovery learning say. And of course mathematics is all around us; so too are atomic physics, gravitation, molecular biology and organic chemistry. They are all, in a sense, though not all in the same sense, 'there'; but the point is that you need more than eyes to see them, and if children are to conceive of their environment in mathematical or scientific ways, they will have to be more than placed in contact with it. They will have to be taught *how* to conceive it.
>
> (1967: 149)

In order to understand how to conceive of the world in a mathematical way, it is necessary to understand something of the activity of doing arithmetic. It seems that, far from arithmetic being abstracted from our actions, it is superimposed upon them; this is a technique that has to be learnt, as an examination of Steffe *et al.*'s (1982, 1983) work in the next section shows.

Understanding the context of doing arithmetic

If knowing about number is a question of knowing how and when to use numbers, then understanding the use of numbers in arithmetic is not a function of having certain essential concepts from which is generated, unproblematically, the ability to do anything with number. On the contrary, understanding the subject matter and methods of arithmetic has to be seen as a matter of initiation into a social tradition. This section deals with the work of Steffe, Thompson, and Richards (1982) and Steffe, von Glaserfield, Richards, and Cobb (1983), whose data on arithmetic problem-solving illustrate that in order to learn how to use numbers in arithmetic, children need to understand the context of doing arithmetic and enter into its associated social practices.

Steffe *et al.*'s view of mathematics is that:

> Acquisition of mathematical knowledge is also a creative activity of the
> knowing subject. In order to be known it must be constructed by the
> individual . . . . The young child, in its state of mathematical naivete,
> but innate curiosity, bears a deep resemblance to the creative
> mathematician working on the frontiers of his discipline. For the child
> also is working on the frontiers of its own knowledge.
>
> (1983: 112)

What Steffe *et al.* overlook is the fact that the 'creative mathematician' has
a large amount of knowledge of what the subject is about and what its
methods are. Mathematicians work within the social context and tradition of
mathematical enquiry. By comparison, the mathematically naive child not
only does not know how to do long division but more importantly does not
know about arithmetic or mathematics: what it is for, what it studies, how
it is used, and what its methods are. This is the social context of
mathematics, which a child has to learn from others.

However, it is clear from what Steffe *et al.* say that they consider
mathematical knowledge to be an individual construction: 'every child must
individually build up an appropriate conceptual structure during the first six
or eight years of life' (1983: 113). Their concern is to identify the nature of
this conceptual development as it relates to the ability to do arithmetic
problems. They thus begin their study by creating a typology of counting
which, they say, is based on what children are aware of when they count;
each counting type – there are five – embodies a particular conception of a
countable unit. The next step is to compare these counting types with the
children's performance on arithmetic word problems: Steffe *et al.* argue that
children's arithmetical abilities can be explained with reference to their counting type. However, a reinterpretation of the evidence suggests that, on the
contrary, the children's performance is dependent on their understanding of
the context of doing arithmetic. Steffe *et al.* assume that the children have
entered into the social practices of doing arithmetic, and in so doing they
overlook the possibility that they were not completely familiar with one
particular social practice: the use of representation in solving arithmetic
problems.

A typology of counting

Steffe *et al.*'s typology is based on 'what it is that the child seems to generate
and be aware of while counting. We call the objects the child creates *unit
items*' (1982: 83). Thus they define counting as 'production of a sequence of
number words such that each number word is accompanied by the production

151

of a unit item' (1983: 116). A total of five unit items are distinguished: perceptual, figural, motor, verbal, and abstract; according to Steffe *et al.*, development follows a cumulative progression from perceptual to abstract. Of interest here are the first three of these; the origins and details of the classification can be found in Steffe *et al.* (1983).

By observing the counting behaviour of a number of children as they performed a task which included counting a known number of items hidden behind a screen, Steffe *et al.* identified: (i) 'counters with perceptual unit items' as children who cannot count unless a collection of items is actually visible: the act of looking and counting is a global event; (ii) 'counters with figural unit items' as children who, on the other hand, can count items that are hidden behind the screen by constructing a figural representation of each item but whose counting actions are restricted to the area of the screen as they count the imagined items; and (iii) 'counters with motor unit items' as children who are able to count the hidden items but are also able to count away from the screen: they too create a figural representation of each item but are aware of the motor act of counting (for example, pointing) as a 'unitary event' and are able to count their actions. However they are still reliant on the actual performance of counting since 'they have not yet fully internalised their counting activity' (1983: 120).

Counters with perceptual unit items

Steffe *et al.* (1982) classified thirty-four 6-year-olds according to counting type and attributed the first three counting types to thirteen of them. Five of these thirteen children were classified as counters with perceptual unit items. They were also considered by Steffe *et al.* to be limited to script-based understanding (Schank and Abelson 1977) of both arithmetical problems and operations: use of a script entails imposing it onto a situation with the result that no further meaning is ascribed to the situation beyond that contained in the script. The scripts for addition and subtraction used by these counters were:

Make an addend corresponding to the first number word heard
Make another addend
Count to find 'how many'

and

Make the minuend
Take something away
Count what's left

(1982: 86)

Steffe *et al.* claim that when the children used these scripts, they did not interpret the actual problems, but activated the relevant script and applied it. Their erroneous solutions were due to their 'perception-bound reasoning' (87), which caused them to consider that referents for number words were created by merely looking at objects. The following protocol is an example:

> **Interviewer:** Bill has three marbles. Tom gives him five more. How many marbles does Bill have now?
> **Chuck:** Would you mind saying that again?
> **Interviewer:** (Repeats problem.) Do you want to use the marbles to help you work the problem?
> **Chuck:** Yeah. (Takes ten marbles from box; looks at the interviewer.)
> **Interviewer:** Bill has three marbles.
> **Chuck:** Okay, three. (Removes three marbles from pile, one at a time.)
> **Interviewer:** Tom gives him five more.
> **Chuck:** (Moves all the remaining marbles to the three he had set aside.) One, two, three, . . . ten. (Counts the marbles in the pile.)
>
> (1982: 87)

According to Steffe *et al.*, Chuck believes that he has constructed the second set of marbles by simply looking at the remaining seven. They support this claim by arguing that such a belief is in keeping with his failure to count the second addend. There is, however, an alternative explanation of his behaviour which is to do with his understanding of arithmetic in general, and the method of representation in particular: he does not see the marbles as an aid to solution of a hypothetical problem – a sophisticated notion in itself – but considers them to be the actual marbles being talked about or at least very closely connected with them. Indeed, Steffe *et al.* have no grounds for assuming that Chuck understands and is familiar with the use of representations of things in the solution of arithmetic problems. If his ordinary activity concerns immediate and real problems, then he has no relevant experience for understanding what arithmetic is about, or what its methods are. In fact, rather than making the problem easier to solve by using actual objects of the type referred to in the problem, Steffe *et al.* may be creating the most difficult conditions possible for a child at this level of contextual understanding, since the possibility for confusion between what is talked about and what is actually present is at a maximum.

This interpretation fits with the fact that Chuck does not count to make a second addend. If he thinks that the marbles in the box are Bill's actual marbles, or are the same as Bill's marbles, then there is no need to count the second addend because all that is required is to count how many marbles Bill has. Thus he counts out three at the prompt of the interviewer, assumes that the remainder are those that Tom gave Bill (he does not need to count them), and counts out the pile, arriving at ten. Alternatively, he may assume

153

that the experimenter has obtained the exact number of marbles in the first place and that there is no need to check; this would be a reasonable assumption, since in elementary arithmetic teaching teachers set up exact representations in the simpler type of problem such as, 'Here are two apples, here are two more apples, how many are there altogether?'. So Chuck's experience of representation may be limited: he does not expect to have to set up a representation for himself, and in the past when representations have been set up they have always been exactly right.

In many of the protocols, the children have to be prompted to use representations of the problem, an indication that they are not familiar with this method of solution. Steffe *et al.* assume that this method is self-evident and that the errors children make lie in an inadequate conceptualization of number. But if a child sees no relationship at all between what is talked about and what is actually present, or if she thinks that they are one and the same thing, or assumes that an exact representation has been set up, then she will behave in a particular way, not because she is rooted in the concrete, perceptual world, but because that is how she must interpret the context. This analysis applies to the following example too which, Steffe *et al.* claim, demonstrates Chuck's belief that looking at a group of objects counts as isolating that group as the set to take away. The question is: 'There are eight buttons in a bag. Jane takes two buttons out of the bag to sew on a dress. How many buttons are left in the bag?'; the square brackets are my addition:

Chuck: (Takes all ten buttons from the objects box.)
Interviewer: There are eight buttons in a bag.
Chuck: Uh-huh (yes). [The interviewer says there are eight.]
Interviewer: How many buttons down there?
Chuck: Eight. [The interviewer just said so.]
Interviewer: Count them.
Chuck: One, two, three . . . ten . . . [Someone is mistaken.]
Interviewer: How are you going to work the problem?
Chuck: You take five away. [He may have forgotten the number involved, or the earlier confusion leads to treating the procedure as arbitrary anyway.]
Interviewer: And?
Chuck: One, two, three . . . five (pointing at each of the remaining buttons).

(1982: 87–8)

Steffe *et al.* comment on Chuck's apparent lack of conflict when he counts ten buttons having said that there were eight. But if Chuck does not understand the context as one in which: (i) the buttons are supposed to represent the ones talked about; (ii) they are not the actual buttons talked about; (iii)

Chuck himself has to set up the initial representation; and finally (iv) the experimenter purposely provides more materials than necessary because this is both a testing and a problem situation in which the answer is not given but is supposed to be – and can be – worked out, then his behaviour at this point is not surprising or irrational. The initial confusion over the number of buttons is never cleared up, and Chuck's probable conclusion is that his counting is unreliable; the result is that finishing the problem becomes an arbitrary process.

Moving on to missing-addend problems, Steffe *et al.* claim that counters with perceptual unit items operate according to the following script:

Make one pile
Make another pile
Count to find how many

(1982: 88)

Thus Beverly:

Interviewer: (Places ten blocks on the table.) Susan had five blocks. Her mother gave her some more blocks, and now she has eight. How many blocks did her mother give her?
Beverly: (Upon hearing 'five' she counted out five blocks, separating them from the pile, re-counted the five blocks, and continued counting the blocks remaining in the original pile.) Ten.

(1982: 88)

Here again Steffe *et al.* analyse Beverly's behaviour as 'making another pile' by simply looking at the remaining pile. Their support for this conclusion lies in the fact that she does not differentiate between ordinary addition and missing-addend problems, and furthermore solves the problem in a random fashion. But again, Steffe *et al.* merely assume that Beverly understands the situation in the first place, that she is familiar with the methods being used and with the purpose of the whole exercise. There is, however, no reason why this should be so.

As in Chuck's case, the interviewer's initial introduction of ten blocks may be confusing, while the syntax of the missing-addend problem compounds the difficulties already present: the point of the problem will elude someone who does not fully understand what doing arithmetic is about and who has limited experience of arithmetic problems. Thus if a child is used to dealing with straightforward questions of the type 'One apple and two apples makes how many apples?' where the unknown is the largest quantity or the net result of the sum, she may consider missing-addend problems to be already answered, because the net result quantity is given, namely, 'Now she has eight'. So the question might be taken to be a statement, not a question at all.

'Her mother gave her some more' is incidental, interest being focused on how many Susan ended up with. Or, fitting the question into the mould of a straightforward addition, a child could treat the 'Now she has eight' statement as a second addend. Either way, she fails to look for and correctly isolate the unknown, a failure which is possibly supported by confusion over the meaning of the ten blocks and what they represent.

The idea that the number of blocks actually present may be understood by a child to have some relation to the number being talked about is supported by Beverly's use of all the blocks. Only by recognizing that these are available so that *some* of them can be used to represent quantities in the problem can she use the blocks successfully. Furthermore, the use of blocks to solve a missing-addend problem is more complex than in straightforward missing-sum problems: in the missing-addend problem it is necessary to count out five blocks, add blocks to reach eight but keep these separate, and then count this second group of blocks. To find the sum of five and three, the second addend (five blocks) does not have to be kept separate. Thus successful solution of the missing-addend problem involves: (i) identifying the unknown and recognizing arithmetic as being about this activity; (ii) understanding representation as a technique; and (iii) understanding various techniques of representation. Lack of any of these can account for Beverly's behaviour.

Counters with figural unit items

Steffe *et al.* (1982) also identified three 'counters with figural unit items'. It is interesting to note that these children did not show any advance over counters with perceptual unit items when they solved problems with objects available: apparently they too had problems with representation. In fact Steffe *et al.*'s theory cannot account for this observation at all, since if the children were, ostensibly, more sophisticated counters, they should not have made the same mistakes as more primitive counters.

Counters with figural unit items used their fingers to represent numbers, a common social practice the purpose of which is probably more obvious than the use of objects to represent numbers. There is one obvious difference between the two methods too, which is that with fingers there is no problem over the ambiguity of how many objects are present to begin with; most people have ten fingers, and they are obviously used in calculation merely as an aid to counting. In the protocols quoted by Steffe *et al.*, it is plain that the children needed a good deal of prompting to use their fingers, but once they did so, they found the problems easy to solve. Before this point, though, they made wild guesses as to the answers, apparently trying to use known number facts. Steffe *et al.* report that these same children had difficulties with missing-addend problems, and showed a tendency to treat them as simple addition. But, as I have already argued with respect to perceptual unit item counters, such a lack of knowledge of arithmetic technique does not merit Steffe *et al.*'s analysis that these children had an inadequate knowledge of numbers.

Counters with motor unit items

Recall that counters with motor unit items are able to execute actions independently of perceptual items because they are aware of the motor act as a discrete experience. Nevertheless, despite the supposedly greater numerical sophistication of these children, Steffe *et al.* (1982) claim that, when given a subtraction problem to solve, 'they seemed to prefer to create an idiosyncratic perceptual referent for the number word for the minuend in the problem statement' (92). Thus:

> **Interviewer:** There are eight buttons in a bag. (Jamie begins taking buttons from the box.)
> **Jamie:** (Takes buttons from box.) There aren't eight buttons here.
> **Interviewer:** Count the buttons.
> **Jamie:** (Puts all buttons in left hand; removes them one by one in synchrony with uttering '1.., 2.., . . . 10').
> **Interviewer:** There are eight buttons in a bag. Jane takes two marbles out of the bag to sew on a dress. (Jamie reaches for buttons.)
> **Jamie:** (Takes two buttons from the pile and places them on cloth.)
> **Interviewer:** How many buttons are left in the bag?
> **Jamie:** (Picks up two from those remaining in pile.) Two. (Continues taking buttons one by one.) Three, four . . . eight. Eight! She has eight left.

And again:

> **Interviewer:** There are eight marbles in a bag. Jane takes two marbles out to play with. How many marbles are left in the bag?
> **Jamie:** How many were in the bag?
> **Interviewer:** Eight. (No marbles were available.)
> **Jamie:** Jane took out two.
> **Interviewer:** (With Jamie.) How many were left in the bag?
> **Jamie:** (Spreads fingers of left hand; counts fingers with index finger of right hand.) One, two . . . five. If you took away two, then I'm gonna have to do it my way! (Counts fingers again, subvocally uttering 1. .2. . . .5; sequentially extends fingers on right hand.) Six, seven, eight . . . Took away two (folds two fingers of left hand). Okay, those two are down. Now let's see (counts remaining fingers). One, two . . . six.

Interviewer: Okay.
Jamie: Okay? Six.

(1982: 92–3)

The first protocol shows clear evidence that Jamie does not fully understand the use of representational items in solving arithmetic problems. The
interviewer never clarifies the situation but instead acts as though there are
eight buttons actually present. (The request to 'count the buttons' in response
to Jamie's statement 'There aren't eight buttons here' is in the nature of a
correction; it implies that if Jamie counts the buttons she will find that there
are eight. When Jamie counts 'ten' the interviewer again states, 'There are
eight buttons in a bag'.) Jamie at this point may simply assume that her
counting is mistaken. This situation is plainly confusing before the child has
even begun to solve the problem, but Steffe *et al.* go on to say that:

> In view of her behaviour in the first protocol, we infer that Jamie did
> create a perceptual referent corresponding to the statement, 'There are
> eight buttons in a bag'. . . . But her perception of the buttons seemed
> to overwhelm any intention she may have had to construct a pile of
> eight. She apparently never counted out a referent for the statement
> 'There are eight buttons in a bag'. . . . Moreover, she experienced no
> apparent conflict in between hearing 'eight' in the problem statement
> and the results of her counting actions. This . . . corroborates the claim
> that she created an idiosyncratic perceptual referent . . . . Her goal
> appeared to be totally satisfied through visual perception of the buttons.

(1982: 93)

Again, Steffe *et al.* completely overlook the problem of understanding
context; they assume that the child is in possession of this understanding, and
consequently overinterpret the protocol. Jamie's successful performance with
no objects present can be explained by her use of the more familiar finger-
counting technique which relieves the situation of many of its ambiguities,
including those attached to the role of the experimenter. Again, Steffe *et al.*'s
counting types classification apparently fails since Jamie's behaviour in the
first protocol shows no greater sophistication with regard to the number
concept. Their theory does not in fact explain all of Jamie's behaviour, and
their observations appear not to relate to children's conceptions of countable
units, but to their understanding of the context of arithmetic, which appears
to be at more or less the same level amongst these 6-year-olds.

Understanding the context of doing arithmetic

Steffe *et al.*'s (1982) protocols lend themselves to a reinterpretation which

158

suggests that, in order to use numbers in arithmetic, it is necessary to enter into the social practices of doing arithmetic. According to this view, the ability to use numbers to calculate is not a matter of understanding certain essential logical principles which then give structure and meaning to any problem. On the contrary, the children in Steffe *et al.*'s study needed to understand a considerable amount about doing arithmetic in order to solve the problems they were given. The analysis here indicates that the social practices of arithmetic include references to the following:

(i) what arithmetic is about: the notion of hypothetical problems; the use of materials as an aid to solution; the nature of problem presentation; the point of missing-addend problems and other difficult syntaxes; the notion of a solvable problem created by someone else as an exercise in finding an unknown;

(ii) methods of representation: the relationship between what is talked about and what is actually present: are the objects present the actual objects in the problem?; the role taken by adults in setting up representation: the distinction between 'materials provided' and an initial exact representation set up by the adult; the rules of setting up representations and subsequent manipulation of objects.

The particular problems shown by Steffe *et al.*'s subjects seemed to be to do with the use of representation in arithmetic, and more generally, how this relates to what arithmetic is about. Learning to do arithmetic is not just a question of mapping a formal code onto informal skills; it is a matter of learning what formal arithmetic is for, what it is about, and what its methods are. Thus arithmetic is a new use of number for the 'mathematically naive child': although formally related to some everyday uses of number and manipulation of objects, it is very different from them. Coming to understand and know how to use number in formal arithmetic must come about through involvement in the context of 'doing arithmetic'.

In this chapter, I have argued that knowing about number is a question of knowing how and when to use numbers. Understanding number use is not, therefore, a function of understanding the concepts of class inclusion, seriation, or one–one correspondence from which is generated, unproblematically, the ability to do anything with number. It is, on the contrary, a question of entering into various social practices, and of coming to share with others certain expectations and interpretations of the contexts of number use. I have already suggested in this chapter what some of the social practices of number use might be, but further analysis is necessary, together with an approach to learning which focuses on the initiation of the individual into those practices. This is the subject of the last chapter.

9

'The development of mathematical thinking': entering into the social practices of number use

The general view in the field of number development and arithmetic research is that there is an inevitable progression towards the understanding of arithmetic in that its basic subject matter is contained within children's minds, needing only abstraction to bring it out. It is assumed that mathematics is a natural product of the human mind. Thus Starkey and Gelman claim that:

> Number forms a natural cognitive domain . . . . Some number abilities are natural human abilities in the sense that some language abilities are natural human abilities . . . . Counting, or more generally the ability to abstract precise numerosities across discrete objects, appears to be a central and basic ability in the number domain.
>
> (1982: 112–13)

In this book I have argued against this view by means of examining theories of number development and related research with regard to two particular problems: the explanation of the growth of number understanding and the description of knowing number.

While I began with a critique of psychological theories of number development which focused on their ability to describe and explain development, I argued, as a result of that critique, that the way in which knowing number is normally characterized is at the root of problems in explaining development; I therefore proposed that knowing number should be reconceptualized as involving entering into the social practices of number use. This has considerable consequences for psychological investigation: it requires the collection of data which can form the basis of an analysis of the social practices of number use and of the way in which children enter into those social practices. This suggests a complete reorientation of the field which rejects the assumption that mathematics is 'everywhere and in everything', waiting to be discovered by the child when he reaches a suitable stage of underlying logic, competence, or concepts. Instead, mathematics must be conceived as knowledge which is intrinsically social and which is constituted

160

by certain social practices: understanding mathematics then involves an initiation into a particular social tradition. It is not a question of 'the development of mathematical thinking' (Ginsburg 1983) in the sense of the individual construction of logical structures necessary and sufficient for the grasp of numerical and mathematical concepts.

In this chapter I aim to show the potential of a reconceptualization of knowing about number as involving entering into social practices by examining the kinds of data and analysis that such a reconceptualization requires. I will illustrate these by drawing on studies of language and learning at home and at school, specifically those conducted by MacLure and French (1981), Walkerdine (1982a–g), Tizard and Hughes (1984), Durkin, Shire, Riem, Crowther, and Rutter (1986), Durkin, Crowther, and Shire (1986), and Durkin, Shire, and Riem (1986). These studies involved extensive recording and/or videotaping of children's conversations with adults and with each other; I will show how such conversations, together with interview and observation of play, can be used as a source of data for the analysis of social practices. Following the original bias of the data, I will focus in the first section on the analysis of the social practices of number use at home and at school. Number use occurs in both these situations within particular social contexts which are distinguished by the intentions and expectations of the participants and the social and personal relations which hold between them; analysis of the social practices of number use must include reference to these wider contexts. In the second section I will examine how studies of the features of word use and topic reference can be used to compare social practices, and how detailed analysis of the features of adult–child interaction can indicate the means by which children are initiated into an adult understanding of situations and the way that words are used in them, and so enter into the social practices of number use.

Analysing social practices

It is worth repeating the claim that mathematics is not 'all around us', waiting to be discovered by the active inquiring mind of the child. More specifically, it follows from this claim that learning mathematics is not a question of simply disembedding mathematical rules and structures from everyday life. While there are links to be made, and arguably they must be made if children are to learn what mathematics is about (Hughes 1983, 1986), the formal links which express isomorphism between everyday action and mathematics are not necessarily self-evident, nor are they sufficient to express the nature of mathematics. An analysis of the social practices of number use in everyday life and in the classroom will show how social practices differ such that number use is a highly differentiated set of activities for which there are no necessary and sufficient conditions.

Drawing the contrasts between different social practices of number use may be more than merely interesting; it may actually be a necessary step in explaining number understanding in that some social practices may only come to light when contrasted with those of another context. This may be particularly true given that social practices involve non-linguistic behaviour such that some assumptions of how to act in a particular situation are taken so much for granted that they are never explicitly commented upon except when someone does not act in accordance with them, but in fact acts according to the social practices of another, related, situation. Much of my reinterpretation of experiments has been based on the corresponding premise that when children fail in a task, they are not simply wrong, but are acting according to another interpretation of the situation, since their lack of familiarity with the present situation precludes the interpretation intended by the experimenter. Not having entered into the social practices of a particular situation, they are at a loss as to how to act, but make the best of it.

A social practice involves a tacit agreement among people in a certain situation as to what constitutes appropriate action in that situation and, reciprocally, what is the definition or meaning of that situation; in this agreement, the meaning of action, including speech acts, and the meaning of the situation, are mutually constitutive. In order to analyse a social practice, it is necessary to state its conventions as they are illustrated by, and embodied in, conversation patterns and discursive styles, the interpretation of particular expressions to refer to particular topics, and the meanings of individual words. Social practices are also indicated by non-verbal behaviour, both as non-verbal communication (gaze, facial expression, physical attitude) and action in response to, or as the initiative for, verbal communication. Related to the rules of conversation, and possibly illustrated by them, are the purposes of the participants within a particular situation, their expectations of each other and of the situation, their assessment of each other's needs and abilities, and their perceived economic, social, and personal relations. It is also relevant to the analysis of social practices to include the historical and social influences on the actors in terms of what motivates their participation in a situation and their subsequent discursive style, and what informs their practice. Finally, one might ask how agreement as to the meaning of a situation and the action in it is arrived at and sustained, how it is changed, and how flexible it is.

Thus an analysis of social practices involves the collection of diverse but related data; much of this may be obtained by observation and interpretation of conversation and action as they occur in particular social contexts. Other relevant data may be gathered from interview, observation of play, and experiment, where appropriate. In this section, I will contrast the social practices of number use at home and at school by examining conversations between adults and children reported by MacLure and French (1981), Walkerdine (1982a–g), and Tizard and Hughes (1984); I will also suggest areas of future research.

Social practices of number use at home

Number use as a highly differentiated set of activities in everyday practices is well-documented in studies such as those conducted by Tizard and Hughes, Walkerdine, and Durkin and his co-workers. Walkerdine classifies activities involving number into instrumental and pedagogic tasks (Corran and Walkerdine 1981); this is a useful distinction which I will follow here. In instrumental tasks, the purpose of the activity is a practical accomplishment or product in which numbers are used to designate items, but do not themselves constitute the main focus of the activity; these form the basis of everyday practice. By contrast, pedagogic tasks focus on numbers themselves, and have as their purpose the testing of knowledge and/or the production of a calculation for its own sake.

Everyday practice

Tizard and Hughes (1984) record many examples of everyday practice involving numbers; these include: playing cards; using numbers in an address to say '*number* six' for a house number; throwing a dice to get '*a* six', and knowing the convention for referring to this; using numbers in time, age, sorting, and classifying (into numbers of big and small objects), fares ('one-and-a-half'), fractions (a quarter/a half of a biscuit), sharing and dividing; and telephone numbers and the conventional way of saying them.

The conversations which Tizard and Hughes report show that, while engaged in these practices with their mothers, the 4-year-old girls in their sample did not necessarily use the numbers correctly nor were they always the ones to use the numbers; often mothers did this. However, the important point is that the girls generally understood the meaning of the overall situation and the purpose of number use in it, responding appropriately to their mothers' use of numbers, and producing numbers in an appropriate, although not necessarily numerically accurate, way themselves. In analysing these everyday practices, it is important to concentrate on the child's understanding of how numbers are used, rather than the number sequence in isolation. This is an extension of Gelman and Gallistel's (1978) observation that children honour counting conventions, even though the produced string is idiosyncratic. The following exchange illustrates one child's understanding of number use in playing knockout whist:

M: (Deals out six cards each.) You have to call. (Child won the previous round, so she decides trumps this time.)
C: Ooh, I got a good hand here again, but I can help it (arranges the cards in her hand). I got two aces here.
M: You shouldn't tell me what you got, go on, call trumps.
C: I call heart, I not putting a heart down.
M: Hearts (mother wins the first trick). That's mine (child wins the

second trick). That's yours (child wins another trick).

C: No, diamonds.

M: Oh, you've changed your mind. That's mine then (child wins the last trick: she has won four tricks to her mother's two).

M: So how many you got?

C: Three.

M: You haven't, count. And I've got . . . ?

C: Two. An' how many did I have?

M: Four. (Child says something unclear: mother deals five cards for the next round.) Five.

C: Oh, I got a good hand here as well again . . . . Ace of spades.

M: What you gonna call?

C: Diamonds.

M: (Looks at child's hand.) You haven't got any diamonds.

C: I have, I got one diamond.

M: You gotta have . . . call one with . . . the highest one. What you got an ace of?

C: Spades.

M: Well call that then.

C: Spades.

(Tizard and Hughes 1984: 52–3)

The use of numbers in knockout whist is fairly complex; it entails recognizing number symbols, knowing which numbers are bigger and smaller than others, counting objects (cards, 'tricks'), and coordinating the whole with the idea of trumps. Another complex use of numbers in everyday practice is that concerning money; Walkerdine (1982a) recorded nursery-school children talking during a lesson in which the focus was on shopping (I will examine this lesson further on pages 174–6):

C1: I've got a teddy bear . . . nine pence.

C2: That's cheap for a bear in't it. . . . How much did you spend Gordon?

C3: Polar bears are usually five hundred thousand pounds.

C4: How much was your basket?

C1: £12 . . . I took it back.

C4: Dear isn't it?

C4: I am rich.

C5: Two pence's not rich. It's not enough to buy bubble gums and a bazooka.

(Walkerdine 1982a: 21–7)

In these examples, the children show how much they understand of practices involving money and what numbers signify within these practices: what constitutes having a lot of money; what is the purchasing power of money; prices as dear or cheap; the idea of spending money; and affordability. Not only do these examples show children using higher numbers than are normally recorded, but they show that they understand what expressions involving number mean in this particular context. Obviously, the children could not count accurately up to five hundred thousand, nor perhaps twelve, and neither might they be able to produce the rule that 100p = £1, but the important point to note is their realistic understanding of numbers in terms of pounds and pence, as indicated by their general talk in this situation. Again, in the following exchange, it is doubtful that the child has much idea of the details of the passing of time in terms of years or even months or weeks, but she can talk meaningfully about age:

C: Do you know, my baby's one now.
T: Your baby's coming here when she's older.
C: She'll go to playgroup when she's two, though.
T: Will she?
C: Yeah. Because when you're two you go to . . . . When I was two I went to a playgroup.
Other C: So did I.
C: That shows you, that people go to playgroup when they're two.
T: Why do they go to a playgroup?
C: Because they're not old enough to go to school.
T: I see. And how old were you when you came here, then?
C: Three or four.
T: Three or four. Then what happens when you're five?
C: You go . . . when I'm five I'll only . . . . I'll go to a . . . . I expect I won't come to here any more.
T: Where will you be, then?
C: Be? In a different school, of course.
T: Do you know which school you're going to?
C: Cross Road (her local primary school).
(Tizard and Hughes 1984: 99–100)

Thus the literature contains many examples of everyday practices involving numbers, counting being the dominant activity according to Tizard and Hughes who report (94) that mothers frequently counted while engaged in everyday activities such as laying the table and making out shopping lists, and asked their children to count in similar activities. Similarly, Durkin, Shire, and Riem (1986) report that mothers consider counting an important part of a child's education, while, for their part, children enjoy counting and often initiate it themselves:

C: Does two and two make . . . four (holds up two fingers of each hand)?
M: Mm.
C: Three and three makes . . . one, two, three, four, five, six (holds up three fingers of each hand and counts them).
M: Mm.
C: Count this . . . one, two, three, four, five, six, seven, eight, (holds up four fingers in each hand and counts).
M: Well done, yes that's right.

(Tizard and Hughes 1984: 54)

Another feature of everyday number use is the use of counting, addition, and subtraction in nursery rhymes, songs, and children's games. Opie and Opie's (1955, 1959, 1969) work shows how prevalent these uses are, and how complex, from the highly rhythmic 'Chinese counting' (1955: 111), to 'The twelve days of Christmas' (1959: 198), to commenting on magpies (1959: 217). Numbers are particularly important in children's games, which do not just involve numbers in the playing, but also in starting the game itself: Opie and Opie (1969: 28ff) report extensively on 'dipping' and again on Chinese counting in this context.

Although children clearly can manipulate numbers in quite complex ways in the contexts of everyday practices and of activity which they initiate themselves (including the use of mathematical terms such as 'makes' in the example above), they also appear to have definite ideas about the limits to their ability to do arithmetic. Thus both Hughes (1981) and Tizard and Hughes (1984: 55) report incidents in which pre-school children claim not to be able to do a task because they do not yet go to school, thus demonstrating their perception of arithmetic practice as closely connected with school learning and as markedly different from everyday number use.

Home pedagogy

The dominance of instrumental uses of number at home does not mean that pedagogic tasks do not occur; they do, but the range of number uses in these tasks is of a particular type: the mothers in Tizard and Hughes' study picked on counting, addition, and subtraction as opportunities for teaching, in contrast to the Piagetian number-related concepts which tend to be the focus of pedagogy in school – one–one correspondence, seriation, sorting, and classifying. This difference is significant given children's performance on tests of these concepts: while pre-school children are used to pedagogic discourse as applied to the 'three Rs', the shift of topic at school may cause considerable problems; I will return to this point in the next section.

Home pedagogy often occurs when the mother turns an instrumental task – in this example deciding how many cakes to make – into an opportunity for teaching:

M: Well, if you're gonna have two, and Daddy's gonna have two, how many's that? (Holds up two fingers, followed by another two fingers.)
C: Three.
M: No.
C: Four.
M: That's right. And if Mummy's gonna have another two, what's that? Four, and two . . . (holds up another two fingers).
C: Four and (unclear).
M: If that's four (holds up four fingers) and that one makes (holds up another finger)?
C: Five.
M: And that one makes (holds up another finger)?
C: Four.
M: No, these are five (points to the five already up).
C: Five.
M: Five and one more is?
C: Six.
M: S'right. Now, if you're gonna have two, and Mummy's gonna have two, and Daddy's gonna have two, that's six, and if Kerry's gonna have two that's, that's six (holds up six fingers again) and one more makes . . . ?
C: Five.
M: No.
C: Six.
M: Yeah, and what's after six?
C: Eight.
M: No.
C: Nine.
M: No.
C: Ten.
M: No.
C: Eight.
M: No, after six, what comes after six?
C: Nine.
M: No, seven.
C: No, not seven!
M: Nine doesn't come after six, definitely seven.
C: My teacher says, 'One, two, then six.'
M: Well you gotta count 'cause you wanna know how many little cakes you gotta make.

(Tizard and Hughes 1984: 175–6)

Not only does the mother use several problem-solving strategies such as breaking down the addition to adding one at a time, using her fingers to represent the cakes, and linking the addition to the known number sequence, she also asks 'pseudo-questions' (Barnes 1969), leaves sentences unfinished for her daughter to complete, checks that her representation is accurate, and finally reintroduces the practical, instrumental purpose of the calculation.

The change in practice is important, and suggests that some links between practices are being made here. Although the mother in this exchange moves into instructional practice, and away from the earlier instrumental exchange, it is important to note that the task has a meaning which she re-emphasizes at the end of the conversation in bringing it back to an instrumental mode. However, as I have argued in Chapter 8, this is not to say that the task is concrete; indeed, in Piagetian terms, it is not, since the cakes under discussion do not exist yet. But they will exist in the not-too-distant future, and it is this aspect of the exchange which gives it meaning. The important point to note here is the mother's switch to a pedagogic style and her daughter's response to it. This response is clearly shown in the following extract:

1. **M:** If you've got three currant buns in the baker's shop (holds up three fingers), look, and I take one away (folds one finger down), how many are left?
2. **C:** (Sings) 'Three currant buns in the baker's shop'.
3. **M:** How many's left if I take one away from three (holds up three fingers and folds one down)?
4. **C:** Two.
5. **M:** That's right.
6. **C:** 'Three . . .' (starts to sing).
7. **M:** No. *Two* currant buns in the shop.
8. **C:** 'Two currant buns in the baker's shop, round and fat with sugar on the top. Come a boy with a penny one day' (stops singing). Put your hand up with two fingers. (Mother does so.) 'Along come a boy with a penny one day' (folds one of mother's fingers down).
9. **M:** How many's left now?
10. **C:** (sings) 'One currant bun in the baker' (stops singing). Put your finger out . . . (Mother holds up one finger) 'Along come a boy . . . and took it away'. None left.
11. **M:** None left now!

(Tizard and Hughes 1984: 94–5)

The child in this extract not only responds to her mother's pedagogic purpose but, at turn 8, herself takes on the role of teacher and instructs her mother on representation by fingers, thus demonstrating her grasp of this general technique and its purpose.

While mothers take the job of teaching seriously as shown in these two examples, they also show flexibility in what they think is worthy of correction and in the ways in which they correct:

M: Do you know any numbers then?

C: One (and starts to count on her fingers).

M: Mm.

C: Two, three, four . . . one, two, three, four . . . one, two, three, four . . .

M: Yeah, what's after four?

C: Five, six, seven, eight, nine, ten.

M: Oh, clever aren't you? What's after ten?

C: One, two, three, four, five, six, seven, eight, nine, ten, eleven, five!

M: Ten, eleven, *twelve.*

C: Ten, eleven, twelve. Fourteen, sixteen, nine, eighteen, sixteen, four, nineteen, seventeen, fifteen, twenty-one, twenty-two, twenty-three, twenty-six, twenty-eight.

M: Oh, that's not bad.

(Tizard and Hughes 1984: 171–2)

The flexibility of mothers in such teaching episodes is illustrative of the general tendency in most conversation to compromise in order to sustain communication. Generally speaking, evidence regarding the nature and frequency of modifications made routinely by adults when they address children can be used as a source of indirect information about a child's intellectual maturity and understanding as assessed by an adult. It can also provide information about how children are inducted into new understanding of words and situations, as suggested by the shift in discursive style in the two teaching examples. One might expect that mothers will be more accurate assessors of their children's abilities than teachers, if only because children at home can be more assertive when they feel pushed, or more open when they do not understand, than they can at school where social constraints operate against this. Tizard and Hughes report that 'the pace was often set by the children' (55) and that they 'protected themselves from any tendency by their mothers to exert excess pressure, by changing the subject or becoming "silly" when too difficult demands were placed on them' (96). Direct information about mothers' assessments of their children's needs and abilities, their reasons for introducing arithmetic skills, and their choice of teaching style will also contribute to the overall picture of pedagogic practice at home and its function. It may provide an illuminating contrast with the same information as provided by teachers.

Social practices of number use at school

Studies of number development and arithmetic skills tend to assume that at school children are simply taught to disembed what they already know. But it appears that school does not do this; rather, it does something much more fundamental: it introduces the child to a completely new social context within which arithmetic teaching takes place. Children enter into the social practices of arithmetic within the larger social situation of the school or nursery school, which is also constituted by distinctive social practices. It is important to remember, though, that arithmetic, in the sense that it involves calculation for its own sake, is not inextricably tied to school: it is also the object of exchanges at home. However, what does appear to be the case is that arithmetic is a pedagogic production; that is to say that, as I have argued in particular in Chapters 7 and 8, arithmetic is constituted by a distinct set of social rules that have to be explicitly stated and learnt in some way which involves a social existence. These rules cannot be simply discovered; so pedagogic practice is an important feature of entering into the social practices of doing arithmetic.

Classroom practice

Nursery teachers seem to be strongly influenced by the Piagetian model of concept learning (Tizard and Hughes 1984; Walkerdine 1982a). Thus Tizard and Hughes report that nursery teachers do not see their educational aims in terms of a curriculum, partly because:

> they see themselves less as teaching, and more as providing the children with a rich learning environment. On further thought, therefore, they will usually say that the selection of play materials constitutes the curriculum. By this they mean that, for example, the provision of vessels of different size and shape for water play helps the child to develop concepts of volume; provision of bricks of different size and weight helps the child who is building towers to recognise causal relationships and develop concepts of balance. In addition, they will usually add that by talking to the child about her play they will help her to learn 'attribute' concepts, for example, size, shape and colour names.
>
> (1984: 181–2)

As these expressed aims suggest, much of the teachers' educational methods involved asking questions, a method suggested by Tough (1976) and termed 'cognitive demands' by Blank (1973); the aim is to 'ask questions which require for their answer the use of definite thinking skills' (Tizard and Hughes 1984: 189) and varying complexities of language use. Cognitive demands are not limited to the classroom; an ordinary question such as 'Why did you go upstairs?' is a cognitive demand as defined here. But teachers' use of cognitive demands tends to involve a testing aspect, not curiosity; the teacher acts with the intention of fostering and assessing the child's 'verbal thinking skills' in asking such 'pseudo-questions'. The following conversation illustrates a variant of the cognitive-demand technique in which the teacher tries to elicit the 'correct' answer from the child:

1. **C:** Can you cut that in half? Cut it in half?
2. **T:** What would you like me to do it with?
3. **C:** Scissors.
4. **T:** With the scissors? (**C** nods.) Well, you go and get them, will you?
5. **C:** Where are they?
6. **T:** Have a look round. (**C** goes over to the cupboard, gets some scissors.) Where do you want me to cut it?
7. **C:** There.
8. **T:** Show me again, 'cause I don't quite know where the cut's got to go. (**C** shows **T** where she wants paper cut.) Down there? (**C** nods; **T** cuts **C**'s piece of paper in half.) How many have you got now?
C: (No reply.)
T: How many have you got?
C: (No reply.)
T: How many pieces of paper have you got?
9. **C:** Two.
10. **T:** Two. What have I done if I've cut it down the middle?
11. **C:** Two pieces.
12. **T:** I've cut it in . . . ? (Wants **C** to say 'half'.)
C: (No reply.)
T: What have I done?
C: (No reply.)
T: Do you know? (**C** shakes head.)
Other C: Two.
T: Yes, I've cut it in two. But . . . I wonder, can you think?
13. **C:** In the middle.
14. **T:** I've cut it in the middle. I've cut it in half! There you are, now you've got two.

(Tizard and Hughes 1984: 194–5)

The teacher does not simply comply with the child's practical request, but sees the opportunity for introducing an educational aspect into the exchange. Her first response at turn two is already testing the child with a pseudo-question, although in asking the teacher to cut (not make, put, do) the paper in half, the child has already displayed knowledge of how it is to be done. In turn eight, she tries to elicit the word 'two', which apparently confuses the child, who, having initiated the whole exchange with a request to cut *one* thing in *half*, may not be thinking along these lines. She supplies the answer readily enough when the teacher makes explicit her question. In turn ten, the teacher pursues the pseudo-questioning further in an attempt to elicit the word 'half', and in so doing rejects the child's correct answer at turn eleven and continues questioning in turn twelve. Rejecting another correct answer (turn thirteen), she finally supplies the answer herself at turn fourteen, using the very construction with which the child initiated the exchange.

The teacher here sets the rules for the nature of the exchange as distinctly pedagogic, and it may be that what the child is in fact learning here are these rules: she is being brought into the social practices of the classroom, and learning in this instance the general nature of interaction with teachers, and, more specifically, the kind of answers they expect. But in this example, the teacher's action cannot be dismissed as mere forgetfulness or worse; to fully understand this aspect of classroom interaction it is necessary to find out exactly what informs the teacher's practice at this point. Tizard and Hughes' comments quoted above give some indication of what this might be.

In some respects classroom practice violates the rules of everyday conversation; Tizard and Hughes report that school conversations are adult-dominated, with the children contributing only 19 per cent (as opposed to 41 per cent at home) of sustaining remarks (for instance, commenting on the other person's contribution, making a spontaneous, but relevant, comment, asking a question). The teacher also dominates by the quantity of her talk: on average she uses fourteen words per turn of talk compared to five words per turn from the children (at home the proportion is more equal at eight words per turn from mothers and six words per turn from the children). The initiators of conversations both at Tizard and Hughes' nursery school and at home tended more than half the time to be children; their purpose was usually to convey information, but the teacher's responses were different from the mothers', and conversations were briefer, adult-dominated, and showed a higher proportion of pseudo-questions. In particular, teachers tended to respond to the child's initial communication with a series of questions:

> **C:** (Playing with clay) I made a seat for the lady.
> **T:** Ooh! That's a good idea. Do you want me to get the others back (referring to other clay models)? Are you going to make another? Do you want to put this lady on it?
> **C:** Yes.
> **T:** Or d'you want to put another one?
> **C:** I want to put that lady on it.
> **T:** Do you want to take the milk jug as well? Do you want to take this as well (the milk jug)?
> **C:** No.
>
> (Tizard and Hughes 1984: 188)

For their part, children rarely asked questions at school, tended to give brief answers, and made little contribution to sustaining the conversation.

One way of looking at this is to say that in fact the children were correctly acting out their particular role in classroom practice: answering questions. Indeed, MacLure and French (1981) observe that children will produce answers even when they do not know them by adopting strategies for guessing what the teacher wants them to say. They also absorb and observe the other conventions of classroom practice involving the question–answer–evaluation sequence (MacLure and French 1981); thus children rarely misinterpret pseudo-questions as genuine requests for information, and the following exchange is unusual:

T: What can you see?
C: And they're going in the sand
T: Mm?
C: You have a look
T: Well you have a look and tell me
I've seen it already
I want to see if you can see

(MacLure and French 1981: 212)

They also learn to respond to the evaluation cues given by the teacher, participating when she invites self-correction by her repetition of the child's wrong answer in the form of a question, or in a flat, unenthusiastic tone (in contrast to the bright tone following a correct response). Note that the meaning of evaluation cues is something that, like the meaning of pseudo-questions, is given by the context; in other situations, repetition of an utterance can be taken to indicate a request for verification, not an implicit evaluation. A child's correct response indicates the extent to which she has entered into classroom practices.

Classroom practice also involves particular turn-taking rules, and again children collaborate in these:

Rights to select next speaker and self-select as next speaker are almost exclusively the teacher's, and when a pupil speaks, it is usually in answer to the teacher. Because all participants collaboratively endorse these asymmetrical rights for turn-taking, teachers retain responsibility.

(MacLure and French 1981: 232)

Indeed, the most important point to note is the collaborative aspect of classroom interaction. As I suggested at the beginning of this chapter, a situation has a particular meaning because the participants in that situation act according to its conventions and thus agree on its meaning and the nature of appropriate action in it (see Edwards and Mercer (1986) for an overview). So in analysing the social practices of the classroom – the context within which most mathematics teaching takes place – it is necessary: (i) to observe its conventions in terms of conversation sequencing and general discursive style; (ii) to consider the influences on the participants: for instance, one might ask what informs the teacher's practice, or what children have been told about school; and (iii) to examine the participants' expectations or intentions: for instance, teachers' reasons for choosing particular topics or practices, and their assessment of children's needs and abilities, or children's expectations of correct behaviour for both child and teacher as evidenced by their general talk and role play. Such data should contribute to some understanding of the way in which children enter into the social practices of the classroom and, within this setting, into the social practices of doing arithmetic.

Doing arithmetic

Arithmetic teaching occurs within the wider context of classroom practice, and involves, as I argued in Chapter 8, a great deal more than merely drawing children's attention to 'the mathematics all around us'. In this section I will illustrate this point by examining two arithmetic lessons reported by Walkerdine (1982a, 1982e), which show how arithmetic as calculation for its own sake involves particular social practices. Analysis of these kinds of data is necessary in order to establish what it is that children know when they know how to use numbers in arithmetic; it shows the potential of the reconceptualization of knowing number as involving entering into social practices, particularly when viewed against the background of classroom practices discussed on pages 170–3.

The first lesson (Walkerdine 1982a) aimed at teaching subtraction through a shopping game in which the children were to 'buy' items for a price less than 10 pence, and calculate the 'change'. Clearly, the idea is to make the task of subtraction 'concrete' by embedding it in the known context of shopping, but the children were not taken in by this, recognizing the important differences between the shopping game and real shopping, the most obvious being that the prices were totally unrealistic (note the children's comments on money on page 164). The most important difference, however, was that what the children were asked to do in the game bore no relation to real shopping, but every relation to classroom mathematics teaching. It is worth listing the aspects of the game and comparing these with real shopping.

The children were given ten plastic 1-pence pieces and told to pick from a pack a card on which was depicted an item (basket, aeroplane, teddy bear, etc.) and its price (less than 10p). They were asked to work out the subtraction operation $10 - x = ?$ with their coins and then record the 'change' from 10p in the form of a written subtraction sum. For each new item that they 'bought', the children started again with 10 pence. The differences between this task and real shopping are numerous, the most important being those to do with the kind of calculation required: shoppers do mental calculation and make estimations, if they bother at all; they certainly do not carry out their calculations with the actual money (real prices would prohibit this exercise, especially if one were dealing with pennies); they do not get another lot of money as soon as they make a purchase; they usually buy more than one item at a time; they do not think of the transaction in terms of 'A take away/subtract B leaves C', but in terms of 'give £A to cover a price of £B and get back £C change'; and perhaps most important of all, the purpose of shopping is to buy things, not to make a calculation.

While the teacher tries to make the task realistic by talking in terms of buying, paying, and change, she also uses the terms 'sum', 'leaves' and 'take away', which are the real object of the exercise. The children played the game but were under no misapprehension as to the real nature of the task, as the protocol reported by Walkerdine shows; the children commented on the cheapness of the items as compared with real prices, and wasted no time on the 'change' calculation, translating the task into arithmetic terms immediately. Just one child took the game too seriously and fell foul of the rules by (quite reasonably) buying more than one item and spending all his money. The teacher has to explain the rules again to him, and here makes very explicit the social practices of the classroom mathematics lesson:

> Each time you get another ten pence to go shopping with . . . . I know what you did, you spent two pence there didn't you and that left you with eight pence and so the next time when your toy cost eight pence you thought you had nothing left, but it says start with ten pence every time doesn't it so you can work out again ten pence take . . . spend eight pence and work out how much you'll have left.
>
> (Walkerdine 1982a: 12)

Notice that this child knows his sums ($10 - 2 - 8 = 0$) and he knows what is involved in shopping; what he did not grasp in this particular task were the rules of this way of teaching, and specifically, the rules of using objects as representations in arithmetic. Notice also the teacher's language: she states the rules explicitly and so indicates how the situation is to be defined. It is significant that she nearly says 'take away', and has to correct herself to say 'spend', thus indicating the particular pedagogic practice that she is engaged in.

One other significant problem that the children had in working through the task was what to do with a blank card:

> C: Miss – if you get nothing on the card what do you do?
> T: What do you think? What do you buy?
> C: You don't write nothing.
> T: So you have to write that sum down. . . . so when you go to the shop you buy nothing . . . so how much money will you have for change?
>
> (1982a: 22)

The use of the blank card seems to push the shopping analogy to its limits such that at the beginning of the exchange neither teacher nor child pretends that this is anything other than a lesson on subtraction, although the teacher later tries to concretize the situation. Despite the confusion over the blank card and what it indicates, the children were generally well aware of what they had to do. The following exchange shows how they recognized the real objectives of the lesson to the extent of ignoring the shopping analogy when doing the calculations, although the teacher is insistent on pursuing the analogy when explaining the game:

> T: and I'm going to take the top card (aeroplane for 5p) . . . and it couldn't be a very big aeroplane could it. So, anyone know what my sum's going to be that I'm going to write down?
> C: Ten take away five leaves five.
> T: Right. And I'm going to use my pennies. I've got ten pence . . . I'm going to spend 1–2–3–4–5 pence buying my aeroplane . . . that's altogether . . . leaves . . .
> C: leaves . . .
> T: . . . leaves me. Gordon? How much does it leave me? How much change? I had five pence change.

> (1982a: 20)

In some respects, the lesson does not appear to teach the children anything they do not already know in terms of subtraction; but it may have confirmed some of the more general social practices of doing arithmetic in terms of the relationship between representation and written calculations, and the production of a calculation for its own sake.

The production of a written representation as an object of arithmetic practice is more clearly shown in the following infant-school lesson reported by Walkerdine (1982e). It may be significant that this object is also clearer to the children than it is in the lesson above. The teacher begins by placing two piles each of two blocks on the table, and immediately focuses by means of pseudo-questions on how many blocks there are. She then goes on to make the blocks into one pile:

> T: How many's there? There? How many's there? Two and put them altogether [sic]. (Makes one pile.) How many have we got already?

The teacher repeats her actions twice, using different numbers of blocks, and then takes another two piles of blocks, three in one and four in the other, giving them to two children. She says:

> T: Good boy, let's count them altogether. One–two–three–four–five–six–seven. So Nicola had four, Debbie had three, so three and four make . . . (she puts the blocks together).
> C: Seven.

> (1982e: 8)

Here the teacher changes the form of what she says to introduce an arithmetical form 'three and four make'; she then begins to move towards a written representation in which two circles are joined by lines to another circle:

She performs the same actions as before, moving the blocks from the top two circles along the lines and into the bottom circle:

> T: and we had four up here and three down there and these mean you've got four and they all go along there and we've got three and they all go along there so we put them altogether and count them and how many have we got? Seven. We've still got seven let's do another one.
>
> (1982e: 9)

The teacher encourages the children to perform the actions and also use the words 'and . . . makes' in synchrony with the actions. She then goes on to remove the blocks and replace them with drawings of blocks:

> T: You've got to trace over first of all everything that's on there. Right? Then you go round the squares and count them as you go. One–two–three–and one and then you've got to draw what you've got there. So we've got to draw three and we've got to draw – are you watching me 'cos you won't know what to do – so I'll put those three in there instead of moving them 'cos we can't pick them up, so I've drawn them in, and three and one make?
> C: Four.
>
> (1982e: 10)

The teacher repeats the actions even though there are no longer any objects, moving her hands down the lines. Notice in her talk her explicit reference to 'what we've got to do', as though she were no longer responsible for the procedure of the lesson, and is referring to some other set of rules concerning doing arithmetic. This is an important point to notice since, despite the teacher's skill in effecting the transfer from blocks to written representation, arguably the children could not follow what was going on without some wider knowledge of doing arithmetic, particularly its methods and purposes, and without some participation in classroom practices.

The next step in the lesson is to move from representations of blocks to written numerals; the teacher gets the children to write the appropriate numeral next to the drawings of the blocks. Again, she refers explicitly to mathematical practice and to classroom practice:

> **T:** Right can you draw a five? Do a five, there. Now in our books just one stage further . . . . Now then in our books. That's how a five goes. Do me a five there. Now then Nicole do you remember you did this at home?
> **N:** Yes.
> **T:** Now this is exactly the same except I've written in the numbers and what you do. Let's do one together first. How many have I written there?
> **C:** Two.
> **T:** And how many have I written there?
> **C:** Two.
> **T:** So then you think, oh I need two – (she gets two blocks) – and I need another two because it says two and then you do exactly the same you move them altogether and count them up . . .
> **C:** It says four.
> **T:** . . . but you don't draw them you write the number. Right, let's do another one. How many does that say?

(1982e: 12)

The children are finally brought to the written representation

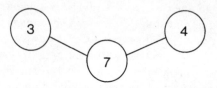

or 'three and four makes seven'.

For Walkerdine, 'the children have been brought to a practice, the object of which is the production of statements of a written form' (12). But it is important not to overlook the more general practices within which this particular lesson takes place and has a meaning; the question remains as to how children enter into those practices which set the general rules of arithmetic discourse and allow the use, for instance, of blocks to represent numbers in the first place.[1] As I have argued in Chapter 8, understanding arithmetic is not a question of the simple disembedding of form from the physical world, and the lessons illustrated here are not merely an exercise in pointing out to the children the mathematics all round them. What is involved is not merely the production of a concept through action on objects, but a much more complex process involving entering into social practices which constitute that action as being of a particular kind.

Entering into social practices

In Chapters 6 to 8 I argued that experiments generated by the Piagetian approach to number development miss the point of what knowing about numbers really involves, producing data which only concern how far children may be said to have entered into the social practices of psychological testing. The reconceptualization of number understanding which I have argued for maintains that knowing about numbers is a question of entering into the social practices of number use, and therefore that development is not a question of internal maturation of a child's conceptual powers as brought about by action in the physical world, but rather of increased social experience and social understanding.

In the last section I aimed to show the complexity and diversity of the social practices of number use and to give some pointers towards analysis. These included the collection of data in three broad areas: (i) the meaning of individual words in particular contexts, the interpretation of expressions to refer to specific topics, and the rules of conversation sequencing; (ii) the purposes and expectations of the participants in a situation, their assessments of each other's needs and abilities, their perceived social and personal relations, and the historical and social influences acting on them; and (iii) the way in which the participants collaborate in, and sustain, the meaning of a situation, and the flexibility of the definition of the situation. In this section I will consider further how data of these kinds can be used to compare social practices and thus suggest how children enter into the social practices of number use. In the first part I will look at studies which compare social practices in terms of word use, and consider some of the implications of these for psychological experiments. In the second part I will consider how a study of adult–child interaction can suggest ways in which children are initiated into an adult understanding of the meanings of situations and of their constituent actions.

Comparison of social practices

I argued in Chapter 6 that word meaning is not fixed, but is an intrinsic feature of social practices; the kind of analysis that is required by a study of changes in word meaning is illustrated by Walkerdine's (1982b, 1982c, 1982d) analysis of relational and size terms. Her analysis focuses on the way in which these relational terms tend to be presented to children with respect to other size and quantity terms, their frequency of production by parent and child, and their use in various situations. What emerges from the analysis is an indication that there are specific and non-arbitrary uses of these terms which are a function of the situation and the actors' intentions, evident in both adult and child speech. A brief summary of Walkerdine's findings here will serve to illustrate the potential for further analysis of this kind.

Walkerdine's analysis tends to concentrate on comparison of contrastive pairs (big/little; tall/short, and so on) across social practices, as there is a suggestion that contrastive pairs formed in arithmetic and psychological test practices may not match those of everyday practices. Thus, in everyday practice, the contrastive pair 'more'/'less', which is so important in conservation experiments, does not often occur. In everyday practice, 'more' tends to predominate in situations to do with food and drink, in which 'less' has little function. In such situations, the opposite of 'more' tends to be 'no more' or 'not as much'. The function of referring to small portions is taken by 'little', which forms a contrastive pair with 'lot'. Equally important to note is the fact that 'same' and 'different' do not always form a contrastive pair; difference is often expressed by use of a dimensional term: 'same' contrasts with 'bigger than'/'smaller than', for instance.

Similar effects in the use of size terms can be noted; as I mentioned in the first section of this chapter, size as a topic is a focus of classroom and testing practices, but not of everyday practices. Furthermore, the use of terms to denote size relations follows different but specific patterns in everyday versus arithmetic and testing practices. Thus 'little' is used significantly more often than 'small' by both children and adults in everyday practice, but 'small' is a word favoured in arithmetic practice. 'Little' apparently has a wider range of application than 'small', and does not just refer to size. Thus 'baby', 'tiny', and 'little' are used synonymously, while 'acting like a baby' is contrasted with being a 'big girl'. Children may also be told that they are 'only little', denoting growing up as well as growing in size, while 'big' and 'little' do not have fixed terms of reference so that a girl may simultaneously be a 'little' girl and a 'big' sister. 'Small' on the other hand is used to refer specifically to size, including the relative fit of objects (clothing, etc.) and sorting and classifying according to size. 'Small' only features when size is the focus of conversation (although this is not the same as size as a topic; in everyday practice 'small' is used when size is significant in achieving a practical outcome).

Walkerdine's analysis of these word uses thus goes some way to explaining children's difficulties in psychological tests and in classroom arithmetic. It makes sense of an infant arithmetic lesson featured in Walkerdine (1982c) in which the teacher unsuccessfully attempted to introduce the term 'middle-size' in the context of a story of the three bears. In so doing, she unwittingly conflated the family relations 'mummy'/'daddy' versus 'baby', otherwise expressed at home as 'big' versus 'little' and including a reference to age as well as size, with the specific 'big' versus 'small' within the context of instructional focus on size as a topic. Given this already complex basis, mismatch in the use of comparatives in instructional/testing practice versus everyday practice seems very likely, as I argued in Chapters 6 and 8.

Another kind of mismatch observed by Pimm (1981, 1987) and Durkin

(Durkin, Crowther, Shire, Riem, and Nash 1985; Durkin, Crowther, and Shire 1986) focuses on the relationship between the number-related lexicon and the everyday lexicon. Thus the number word 'one' is also used as a deictic pronoun, while the numbers 'two' and 'four' have homophones (to, too, for). More broadly, Durkin *et al.* (1985, 1986) point out the mismatch which occurs through the existence of 'polysemous' vocabulary. Thus several everyday spatial terms can also be used in mathematical contexts – for instance numbers go 'up' or 'down', some numbers are 'high' and others are 'low', seven is 'below' fourteen but fourteen is 'above' seven and 'next' to fifteen, we can have 'big' numbers and 'small' numbers and finally numbers 'rise' and 'fall'. Given that the spatial sense of these terms is that which occurs in everyday practice, we might expect children to experience some difficulty in dealing with such vocabulary as it occurs in mathematical contexts and, indeed, Durkin *et al.* (1985, 1986) report that this is so.

As these analyses show, the use and understanding of individual words is dependent on their appearance in particular contexts, and this dependence takes two forms: (i) word use changes according to context in terms of the relations between words, that is, in terms of the contrastive pairs which they form; and (ii) word use changes according to the context in terms of the general topic referred to – for instance, 'small' as it is used in everyday practice refers to practical topics such as the fit of clothing, but in arithmetic practice 'small' refers to the topic of size *per se*. These differences indicate the care which must be taken in experimental design to ensure that both experimenter and child share the same topic reference; as Freeman *et al.* (1982) point out with reference to Donaldson and Lloyd's cars and garages experiment which I discussed in Chapter 8, 'whenever children give strange answers in such studies, it is up to the psychologist to *prove* that he or she has really established joint reference with the child in the way that was intended' (1982: 70). Establishing such a joint frame of reference in psychological tests may be very difficult given that an important feature of the situation is the fact that it is a psychological test with an asymmetrical power relationship (Freeman, Sinha, and Condliffe 1981). Children's lack of understanding of the practices of tests and puzzles, and their inferior status in the test situation must contribute to considerable difficulties if topic-setting cues are to be controlled.

There is, then, no 'standard' perception of a task.[2] As I argued in Chapter 8, producing the correct answer in, say, a Piagetian test entails an appreciation of public standards as to what constitutes appropriate behaviour in such situations. These considerations draw attention to the wider issues of social interaction involved in entering into the social practices of number use: these concern the way in which the participants in a situation arrive at an agreement as to its meaning and the meaning of actions within it, and how this meaning is sustained or changed. The way in which some of these issues might be studied is the subject of the final section.

Learning through interaction

In the first part of this section I discussed studies which show how social practices differ according to word use and topic reference, and how this comparative data can be analysed to explain children's apparent failure in arithmetic lessons and psychological tests in terms of their misunderstanding of adults' intentions. Given this analysis one might then consider the processes by which children are initiated into that adult understanding of the meaning of situations and their constituent words and actions. Some indication of the nature of these processes was implicit in my discussion of social practices in the first section of this chapter; for instance, I remarked on certain features of mother–child and teacher–pupil exchanges concerning styles of teaching and correction and children's responses to these. In this final section I will consider these features in greater detail with respect to the way in which children enter into the social practices of arithmetic.

An important feature of conversation is that it is sustained by the participants and that they monitor each other's talk in order to produce appropriate next moves and keep the conversation going. When conversation involves adults and children, however, it has special features relating to the fact that children do not always share an adult's understanding of the situation, and therefore that both participants have to make special efforts to sustain the conversation in terms of shared meaning. Thus when adults talk to children, they make special efforts to comprehend by constantly requesting clarification of utterances, and they also modify their own speech according to their assessment of the child's needs; children for their part respond to invitations to self-correct and to overt corrections, and take part in patterns of interaction set up by the adult. So, for instance, adults may formulate a conversation by summarizing or translating it or commenting on its accordance with or departure from certain rules (Garfinkel and Sacks 1970: 350); they may also reshape a child's utterance in terms of lexis or syntax, for example:

C: I don't see some more
A: You can't see any more

(Durkin, Shire, and Riem 1986: 7)

More specifically, Durkin (Durkin, Shire, Riem, Crowther, and Rutter 1986) identifies three number-related pedagogical strategies used by mothers of under-3-year-olds. These are: (i) repetition and clarification of cardinality (breaking the input down into component parts):

M (to Alan, 18 months): Put two in. That's one, that's one.

(279)

(ii) recitation of number strings:

M (to Edward, 22 months): How many have you got?
C: One
M: One two, one two three four

(280)

and (iii) alternating number strings (prompting the child to give the next word in the sequence):

C: One
M: Two
C: One
M: No, two

(281)

In spite of these strategies, though, Durkin, Shire, and Riem (1986) and Durkin, Shire, Riem, Crowther, and Rutter (1986) report that number words are a source of considerable potential confusion for children, not only because of the degree of lexical ambiguity associated with them due to double meanings, homophones, and so on, but also because parental input frequently violates the conventional number order. Thus, Durkin and his colleagues argue, the linguistic input is potentially very confusing and, in fact, 'we found very few instances where joint verbalising about numbers seemed to run smoothly' (Durkin, Shire, and Riem 1986: 5):

C: I was playing with lots of things
A: Like what? [. . . .]
C: Hundred of 'menies'
A: What?
C: Hundreds of 'menies'
A: Hundred of 'menies' what's that? What's hundreds of 'menies'?
C: Yer. That mean 'dunda'
A: What?
C: That mean 'dunda'
A: 'Dunda'
C: Yer
A: What's 'dunda'?
C: It it means 'zunda'
A: 'Zunda'
C: Yer
A: Are you talking nonsense? Are you?
C: (*laugh*) Yer

(Durkin, Shire, and Riem 1986: 8)

But much interaction does not involve such explicit attempts to use and understand words correctly. Durkin, Shire, Riem, Crowther, and Rutter (1986) record a number of conversations like the following:

> **M** (to Edward, 36 months): How many leaves are on that plant this week?
> **C:** They are hundred
> **M:** There isn't. *Can you*
> **C:** *Can e*
> **M:** Count them?
> **C:** Shall I count them?
> **M:** Yeah you count them, see how many there are. See if you can get it right
> **C:** One, two, three, four, five, six, seven, eight, nine, ten, eleven, twelve, thirteen, fourteen, sixteen, eighteen, nineteen, twenty, eighteen, (points to one leaf on each count word, but counts some leaves more than once, and emphasizes the final word.)
> **M:** That's a lot isn't it?
> **C:** Yea
> **M:** (laughs) I think you counted some of them twice. Didn't you?
> **C:** I think I didn't
> **M:** I think you did
> **C:** I fink I didn't. One, two, three, four, five, six, seven, eight, nine, ten, eleven, twelve (pointing to a new leaf with each number)
> (Durkin, Shire, Riem, Crowther, and Rutter 1986: 285)

Durkin *et al.* note that while Edward is keen to show his counting abilities, his mother's frequent challenges over accuracy eventually have some effect: 'Edward refuses to play according to the rules . . . nevertheless, in the course of the game he receives considerable guidance about the adequacy of his counting and does modify his behaviour in response' (286). As I argued in Chapter 6, children may learn through their awareness of adult responses to their actions and to the words they use, and thus are motivated to act more appropriately in order to be able to communicate more successfully; correction may take the form of *self*-correction by means of, as Hamlyn says, doing things 'as *they* do' (1978: 102).

The importance of establishing a shared understanding of the meaning of a situation and its constituent actions is particularly apparent in arithmetic teaching. As I stated earlier (on pages 169–78), the understanding of arithmetic is a pedagogic production: it is not a matter of the individual discovery of mathematical relations through the experience of action on objects, or cooking, shopping, or playing cards, but requires the explicit statement of rules and conventions on the part of the teacher in order to produce a shift of meaning of words and actions into that constituted by their occurrence in arithmetic. Thus it is necessary to understand what a teacher does in order to produce that shift, and how children respond to the teacher's redefinition of the situation.

There are important parallels to note between what teachers do in arithmetic classrooms and what adults do in their everyday interaction with children. Just as adults in everyday conversations formulate and reshape children's utterances to fit with adult ways of talking, teachers reformulate children's references to number to suit the purpose of the lesson, and they redefine the meanings of actions to fit with an arithmetical interpretation of the situation. In the following extracts reported by Corran and Walkerdine (1981), the teacher is concentrating on place value and the reformulation of number as tens and units, which she has been presenting to the children in terms of bundles of matchsticks and single matchsticks:

Michael: . . . forty then.
Teacher: Just forty, just four bundles of ten, right.

Or, more collaboratively:

Michael: Twelve.
Teacher: Twelve yes. – (She points to the '1' of the '12') – how many bundles of tens?
Children: One.
Teacher: (Pointing to the '2' of the '12') – And how many single ones?
Children: Two.
Teacher: Two right.

Or using prompts:

Teacher: (She points to the '2' of Ahmed's '24'). Two and four.
Children: Ten.
Teacher: (Pointing to the '4' of the '24'). And four . . . ?
Sue: Single ones.
(Corran and Walkerdine 1981: 77)

In these examples, the children could not respond to the bundles and singles questions without some understanding of the relation of representation between the bundles of matchsticks and the numbers being talked about. Earlier in the lesson, the teacher establishes this relation:

Teacher: Seventy? – (As the teacher says this she writes '7'.)
Children: Six.
Teacher: Six – (As she writes the '6' to complete '76'.) – If we were

185

talking about Anne's matchsticks what would that seven mean? (The teacher points to the '7' of the '76'.) Think about Anne's matchsticks that she put out for us. Anne what does seven mean (she points to the '7' again) when I'm talking about your matchsticks? What have you got – seven . . . ?

Anne: Bundles
Teacher: Seven bundles of . . . ?
Anne: Ten

(1981: 79)

Although the teacher intends to make place value 'concrete' by embedding it in the activity of putting matchsticks into bundles of ten, her lesson can be seen as one in which the rules of place-value are used to give a particular meaning to the activity, and the explicit statement of the representative role of the matchsticks is an important feature of the lesson. Had she not done this, there would be no reason for the children to attach that particular significance to the activity; it is not self-evident. Significantly, the children see the grouping into tens as arbitrary, and the teacher has to explain it by appeal to convention:

Teacher: Do you think it could have been eleven or twelve in the bundles?
Children: Yes
Teacher: No, but you can count in tens and use one that on the abacus don't we when we're counting (sic).

(1981: 87)

These extracts underline the fact that the children are not learning by discovery; the teacher explicitly relates the activity of putting matchsticks into bundles of ten to the rules of place value which she is trying to teach and thus draws the children's attention to the features of the situation that she wants them to notice and learn to describe in a particular way. But in so doing, she is not immediately bringing about the intended shift; the children cannot come to interpret the situation as the teacher does by passively learning new words for numbers or a new description of groups of matchsticks: they must also participate in the construction of meaning and respond actively to the teacher's redefinition, in the same way that they do in their everyday conversation with adults. Thus they are prompted by the teacher into the reformulation of numbers as tens and units, and they enter into a dialogue with her, although as the less knowledgeable participants. Coming to share her interpretation of the situation will then be at least in part a result of their collaboration in classroom practice and its particular patterns of interaction; learning to act appropriately in response to the teacher's questions will be a first step in entering into arithmetic practices.

In earlier chapters, I have shown how the social dimension of knowledge has been largely ignored due to the dominance of the Piagetian paradigm within the field of child development; this failure to acknowledge the intrinsically social nature of understanding has led to insurmountable problems of description and explanation, not only in Piaget's original theory, but in those of his rivals and his successors. This chapter represents a first step in a reorientation of the field which describes and explains the growth of understanding and the nature of understanding itself in terms of the acquisition of social knowledge within the context of a social existence. In it I hope to have shown the potential of a reconceptualization of knowing number as entering into social practices.

Notes

Chapter One The development of the number concept as a field of psychological investigation

1. I should point out here that I intend there to be no difference in meaning between 'knowing about numbers', 'knowing numbers', 'knowing how to use numbers', 'knowing the meaning of numbers', and 'number understanding'.

Chapter Three The child's conception of number

1. This feature is particularly evident in Brainerd's (1979) theory, which has its basis in the argument that Piaget's synthesis of order and class is incorrect, and that psychological data support the view that seriation is the most primitive notion in the development of the number concept.

Chapter Five Does Piaget give an adequate account of growth?

1. There may be some question as to the application to Piaget's theory of Fodor's description of concept learning as the projection and testing of hypotheses: it does not seem wholly accurate although this may be because Piaget tends to concentrate on the acquisition of quantifiers such as 'all' and 'some' (to which Fodor's last example best applies), rather than on the acquisition of concepts that can be expressed in terms of predicates, such as 'dog'. However, such concepts do appear in Piaget's account as Chapter 4 showed: they provide the vehicle for development of the concepts 'all' and 'some'.

2. Smith (1982a, 1984a, 1984b) argues that Piaget is invulnerable to Hamlyn's criticism since Piaget does not purport to state both necessary *and* sufficient conditions for the acquisition of knowledge (1982a: 180). This is false, however, since although Piaget may claim to state only necessary conditions for the acquisition of knowledge, the fact that he claims to state both necessary and sufficient conditions for the possession of knowledge (see my argument in Chapter 6, pp. 80–9) has repercussions on his picture of acquisition: the deficiencies in Piaget's account of knowing show that his account of coming to know fails to express a fundamental requirement of the growth of knowledge which, as Hamlyn argues, is a social existence.

Relatedly, Smith argues that Hamlyn assumes the 'adult knower' and thus misses the point of Piaget's attempt to characterize children's knowledge: Hamlyn (1983) argues that: (i) knowing something as true presupposes (ii) knowing what it is for something to be true, which (iii) involves the appreciation of the force of a norm which (iv) implies correction by others which is seen as correction, which (v) implies seeing the source of correction as a corrector with intentions, desires, and interests; this argument is misread by Smith (1982a: 177) to mean that a child must understand the logical connections between these points, and he claims that Hamlyn's account is wrong since young children cannot make such logical connections. But Smith's defence of Piaget rests on a considerable misconstrual: Hamlyn does not intend to argue that points (i) to (v) are logical steps proceeded through by a developing child;

rather, he intends to point out why a context of personal relations must be present for a child to gain knowledge. Smith has confused an argument about a child's knowledge/thinking with the content of that knowledge/thinking.

3. Moore (1980) criticizes Hamlyn for ignoring 'the other sense of "right"', which is the 'non-social one of conformity to, or recognition of, the way things are, where the force of an independent world is felt' (1980: 261). Moore asserts that appreciating something as true or correct can initially occur through contact with the natural world, and thus that a child does not have to be taught this by others: for instance, a child who tries to reach an object but fails a number of times 'believes' that he cannot reach it and so can be said to have 'a true belief and good reason for the belief'. As a result, he 'knows what it is to be mistaken [and he] *knows it as true* that he cannot reach the object' (262); he does need to conform to a social standard of correctness or to appreciate the force of a norm. Furthermore, she argues, a child could not appreciate the force of interpersonal correction unless 'he was independently capable of appreciating what "being right" was all about' (262). However, Moore assumes that it is possible to attribute both beliefs and the concept of an object to an infant, but as Chapter 4 shows, this is problematic; in his reply Hamlyn (1981) argues that, while Moore may be right in arguing that a relation to the natural world is necessary if one is to appreciate something as true, it cannot be shown to be sufficient.

Chapter Six Do number theorists give adequate accounts of knowing?

1. It is appropriate to add here a note on Brainerd's (1979) alternative 'ordinal theory' which, he claims, makes better logical sense than Piaget's synthesis of order and class. The theory assumes that 'number refers to the terms of transitive-asymmetrical relations and that the natural numbers are . . . devices for representing the terms of the progressions which such relations generate' (1979: 100). It proposes that examples of the logical idea of progression exist in the physical environment and that people perceive the transitive-asymmetrical relations which underlie them. Such perceived ordinality Brainerd argues to be the pre-numerical basis of the number concept. Brainerd's account of the development of the number concept is similar to Piaget's even though it is critical of it. Not only does Brainerd use the mechanisms of abstraction and internalization to account for development, but his treatment of knowing about number also assumes that it is possible to state the necessary and sufficient conditions for being said to know number. Instead of Piaget's synthesis of order and class, Brainerd argues that understanding transitive-asymmetrical relations is the necessary and sufficient condition for possession of the number concept. For similarly essentialist theories of arithmetic ability, see also Carpenter and Moser (1982) and Riley *et al.* (1983).

2. My argument here is very similar to Hamlyn's, although he argues further that knowing something is knowing it as true, which in turn 'presupposes possession of the concept of truth or something like it to some degree or other'(1983: 164). Hamlyn goes on to add that this involves not only the appreciation of the force of a norm, but the concept of a norm. Entering into social standards of correctness, he argues, is a matter of being corrected by others, and of seeing that as correction; the source of that correction must be seen as a corrector with intentions, desires and other qualities of personhood. Cooper (1980) criticizes Hamlyn's argument on the grounds that it treats the move from 'x knows that p' to 'x knows that p is true' as trivial, whereas, to generate the rest of the argument, he must include the concept of truth. In fact, examination of Hamlyn's argument shows that he does include the concept

of truth: he argues that the concept of truth, the concept of a norm, and the concept of correctness ('seeing correctness as correctness') are necessary in knowing that something is true or correct. But to talk of possession of the concept of truth in this apparently strong sense of seeing truth as truth raises the problem of the origins of such possession. This point is also raised by Elliott (1980) and Jones (1981), who argue that Hamlyn's account of concept possession leads to the attribution of innate knowledge to children. Elliott also criticizes the 'degree of reflectiveness' in Hamlyn's account of knowledge; he argues that possession of a concept of truth is not necessary for this concept to be operative in one's life.

3. Bryant and Trabasso (1971) criticized the design of Piaget's transitivity experiments from the point of view that they made too heavy a demand on young children's memories and also allowed them to arrive at a non-logical solution by simply remembering the appropriate 'larger than' and 'smaller than' labels attached to A and C respectively in the standard 'A is larger than B and C is smaller than B' comparisons. By redesigning the transitivity problem to ensure that children remembered the original premises and could not respond with a simple reproduction of labels, Bryant and Trabasso found that 4-, 5-, and 6-year-olds were able to solve the problem and concluded that they could make transitive inferences.

4. Gelman (1982) retracts the claim that very young children are totally unable to reason about unspecified numerosities: she accepts Bryant's argument that they use one–one correspondence to judge equivalence and also remarks that, if children can be got to produce correct conservation judgements as in Markman's (1979) study, then they must be able to access 'a quantitative use of the principle of one–one correspondence' (1982: 211). Gelman suggests that pre-schoolers' use of the one–one principle in this study is based on their ability to count, which involves a kind of one–one correspondence in the sense that a unique tag must be assigned to each element in a collection; although this does not require an explicit use of the one–one principle, its understanding is implicit. Thus Gelman argues that pre-school children fail to conserve number because they lack an explicit understanding of one–one correspondence: they have an implicit understanding but are unable to put it into operation. Accordingly, Gelman trained children to notice that displays which are in one–one correspondence have the same cardinal number but those which are not have different cardinal values. Their performance improved considerably in conservation post-tests with set sizes too big to count, and Gelman concludes that they accessed an otherwise implicit ability to use one–one correspondence. However, although this paper represents a change in Gelman's views regarding children's early abilities, it does not represent any change in the substance of her theory in terms of what is involved in either counting or the ability to conserve.

Chapter Seven   Can a Piagetian perspective be defended?

1. Glachan and Light (1982) present an explanation of these results in terms of differences in the resolution of conflict between tasks requiring verbal explanation and tasks requiring manipulation of materials. Note that they retain the notion of conflict resolution, however, thus failing to address the issue here. Similarly Light (1983: 74) suggests that an important distinction between the tasks is that spatial coordination is a 'problem-solving task', whereas conservation is a 'judgement task'.

2. See Wilkinson (1976) and Trabasso *et al.* (1978) for detailed task analyses of class inclusion.

3. Flavell (1978) argues that Case does not give clear details of the way in which the demands made on working memory by any particular task are to be calculated.

The result of this is that Case's cross-task and cross-strategy comparisons in terms of memory demands tend to beg the question. Flavell (1978: 101) argues further that Case's theory leaves other questions unanswered, for instance how working memory and its limits or capacities should be conceptualized; whether children would in fact make faster progress from less to more memory-demanding strategies if a strategy's memory demands are artificially reduced; and how to decide that a mental operation is 'sufficiently automatized'.

4. De Corte and Verschaffel's (1981) analysis of arithmetic instruction also demonstrates the necessity for learning about the context of arithmetic in order to be successful in solving problems. They claim that the errors children make when attempting to solve more complex problems are due to an inadequate conceptualization of the part–whole relation and the equality concept. However, close inspection of De Corte and Verschaffel's experimental teaching programme, which was designed to teach these two concepts, suggests that the improvement shown by the children was due to the fact that the programme made explicit a set of skills and methods in arithmetic which enabled them to enter further into its social practices: the children were familiarized with arithmetic symbolism, with the roles of givens and unknowns and their identification, with the function of representation as an aid to solving problems, and with a means of checking the answer for acccuracy.

Chapter Eight Knowing how and when to use numbers

1. Although Miller (1977) failed to replicate this finding, Silverman (1979) did succeed in doing so, and also pointed out that Miller's results might be due to the fact that his subjects were tested over a series of different conservation tasks, leading to possible carry-over effects. Silverman also points out that Miller's (1976) failure to find increased conservation performance in a non-verbal task in which children were asked to choose which of two rows of sweets they would like to eat may have been due to the fact that the question was repeated both before and after the transformation.

2. This point is also relevant with respect to Fluck and Hewison's (1979) finding that when the conservation task was presented by an adult on television, children's responses did not improve in comparison with the standard task, even though the adult who asked the question was not the one who manipulated the materials. Fluck and Hewison did, however, find significant improvement when the task involved televised puppets arguing over the results of the conservation transformation. In contrast to the standard situation, the puppet argument may present a clear and more natural focus on number and indicate to the child that a numerical judgement is called for in response to the test question.

3. See Sinha and Walkerdine (1978) and Walkerdine and Sinha (1978) for a discussion of this which suggests that non-conservation responses are also a result of the use of 'more' to refer to one dimension only, not two.

Chapter Nine 'The development of mathematical thinking': entering into the social practices of number use

1. According to Walkerdine (1982g: 137–8) a statement can be described either in terms of its internal relations of combination (the metonymic axis), or in terms of the relations of selection of its elements, the source of which is external to the

relations inside the statement itself (the metaphoric axis). She thus contrasts context-dependent 'practical reasoning', which judges the truth of a statement in terms of its correspondence to the rules of a practice, with 'formal reasoning', where truth is determined with respect to the internal relations of the statement itself; such reasoning entails ignoring the metaphoric content of a statement. For Walkerdine, the metaphoric content of mathematical discourse is suppressed such that, 'the task then becomes one of explaining how children learn to reflect consciously on their practices and, later, on the internal relations of the statements themselves' (138). Here, I would argue, it is important to retain the idea of mathematical reasoning as a distinct social practice with its own discourse and use of metaphor; arithmetical statements are not merely 'disembedded' descriptions of everyday practices, and arithmetical metaphor (for example using matchsticks to represent 'tens' and 'units') is not self-evident; therefore seeing the metaphor as metaphor is part of the social practice of doing arithmetic.

2. This fact is well demonstrated by Walkerdine and Sinha's (1978) 'locative orientation test' which demonstrated that both children and adults could construe a task of placing a toy vehicle in front of or behind another in a number of ways according to the degree to which they assumed the task to refer to actual road use.

References

Barker, S. (1964) *The Philosophy of Mathematics*, New Jersey: Prentice Hall.

Barnes, D. (1969) *Language, the Learner and the School*, Harmondsworth: Penguin.

Beth, E.W. and Piaget, J. (1966) *Mathematical Epistemology and Psychology*, Holland: Reidel.

Beveridge, M. (1982) *Children Thinking Through Language*, London: Edward Arnold.

Blank, M. (1973) *Teaching Learning in the Pre-School*, Columbus: Charles E. Merrill.

Brainerd, C. (1979) *The Origins of the Number Concept*, New York: Praeger.

Bruner, J., Olver, R., Greenfield, P., with Hornsby, J.R., Kenney, H., Maccoby, M., Modiano, N., Mosher, F.A., Olson, D.R., Potter, M.C., Reich, L.C., and McKinnon Sonstroem, A. (1966) *Studies in Cognitive Growth*, New York: Wiley.

Bryant, P. (1972) 'The understanding of invariance by very young children', *Canadian Journal of Psychology* 20: 78–9.

Bryant, P. (1974) *Perception and Understanding in Young Children*, London: Methuen.

Bryant, P. and Trabasso, T. (1971) 'Transitive inferences and memory in young children', *Nature* 232: 456–8.

Cantor, G. (1895–7) 'Beiträge zur Begründung der transfiniten Mengenlehre', *Mathematische Annalen* 46: 481–512; and 49: 207–46, trans. Ph. E.B. Jourdain in *Contributions to the Founding of the Theory of Transfinite Numbers*, Chicago and London: Open Court, 1915.

Cantor, G. (1899) 'Letter to Dedekind', in J. van Heijenoort (ed.) *From Frege to Gödel, A Source Book in Mathematical Logic, 1879–1931*, Cambridge, Mass.: Harvard University Press, 1967.

Carpenter, T. and Moser, J. (1982) 'The development of addition and subtraction problem-solving skills', in T. Carpenter, J. Moser, and T. Romberg (eds) *Addition and Subtraction: a Developmental Perspective*, New Jersey: Lawrence Erlbaum Associates.

Case, R. (1978a) 'Piaget and beyond: toward a developmentally based theory and technology of instruction', in R. Glaser (ed.) *Advances in Instructional Psychology (vol. I)*, New Jersey: Lawrence Erlbaum Associates.

Case, R. (1978b) 'Intellectual development from birth to adulthood: a neo-Piagetian perspective', in R. Siegler (ed.) *Children's Thinking: What Develops?*, New Jersey: Lawrence Erlbaum Associates.

Case, R. (1982) 'General developmental influences on the acquisition of elementary concepts and algorithms in arithmetic', in T. Carpenter, J. Moser, and T. Romberg (eds) *Addition and Subtraction: A Developmental Perspective*, New Jersey: Lawrence Erlbaum Associates.

Case, R. (1985) *Intellectual Development: Birth to Adulthood*, New York: Academic.

Chomsky, N. (1957) *Syntactic Structures*, The Hague: Mouton.

Chomsky, N. (1959) 'Review of *Verbal Behaviour* by B.F. Skinner (Appleton-Century-Crofts, 1957)', *Language* 35: 26–58.

Chomsky, N. (1965) *Aspects of the Theory of Syntax*, Cambridge, Mass.: MIT Press.

Chomsky, N. (1968) *Language and Mind*, New York: Harcourt Brace.

Cooper, D.E. (1980) 'Experience and the growth of understanding', *Journal of Philosophy of Education* 14: 97–103.

Corran, G. and Walkerdine, V. (1981) *The Practice of Reason Vol I: Reading the Signs of Mathematics*, University of London Institute of Education Mimeo.

References

De Corte, E. and Verschaffel, L. (1981) 'Children's solution processes in elementary arithmetic problems: analysis and improvement', *Journal of Educational Psychology* 73, 6: 765–79.

Dearden, R.F. (1967) 'Instruction and learning by discovery', in R.S. Peters (ed.) *The Concept of Education*, London: Routledge & Kegan Paul.

Doise, W. and Mugny, G. (1975) 'Recherches sociogénétiques sur la coordination d'actions interdépendantes', *Revue Suisse de Psychologie* 34: 160–74.

Doise, W. and Mugny, G. (1979) 'Individual and collective conflicts of centrations in cognitive development', *European Journal of Psychology* 9: 105–8.

Doise, W. and Mugny, G. (1984) *The Social Development of the Intellect*, Oxford: Pergamon.

Doise, W., Mugny, G., and Perret-Clermont, A-N. (1975) 'Social interaction and the development of cognitive operations' *European Journal of Social Psychology* 5, 3: 367–83.

Doise, W., Rijsman, J., Van Meel, J., *et al.* (1981) 'Sociale marketing en cognitieve ontwikkeling', *Pedagogische Studien* 58: 241–8.

Donaldson, M. (1978) *Children's Minds*, London: Fontana.

Donaldson, M. and Balfour, G. (1968) 'Less is more: a study of language comprehension in children', *British Journal of Psychology* 59: 361–471.

Donaldson, M. and Lloyd, P. (1974) 'Sentences and situations: children's judgements of match and mismatch', in F. Bresson (ed.) *Problèmes actuels en psycholinguistique*, Paris: Centre Nationale de la Recherche Scientifique.

Donaldson, M. and Wales, R. (1970) 'On the acquisition of some relational terms', in J.R. Hayes (ed.) *Cognition and the Development of Language*, New York: Wiley.

Durkin, K., Crowther, R., and Shire, B. (1986) 'Children's processing of polysemous vocabulary in school', in K. Durkin (ed.) *Language Development in the School Years*, London: Croom Helm.

Durkin, K., Crowther, R., Shire, B., Riem, R., and Nash, P. (1985) 'Polysemy in mathematical and musical education', *Applied Linguistics* 6, 2: 147–61.

Durkin, K., Shire, B., and Riem, R. (1986) 'Counting on you but depending on me: parent–child interaction and number word acquisition', in R. Crawley, R. Stevenson, and M. Tallerman (eds) *Proceedings of the Child Language Seminar*, Durham: University of Durham.

Durkin, K., Shire, B., Riem, R., Crowther, R., and Rutter, D. (1986) 'The social and linguistic context of early number word use', *British Journal of Developmental Psychology* 4: 269–88.

Edwards, D. and Mercer, N. (1986) 'Context and continuity: classroom discourse and the development of shared knowledge', in K. Durkin (ed.) *Language Development in the School Years*, London: Croom Helm.

Elliott, R.K. (1980) 'D.W. Hamlyn on knowledge and the beginnings of understanding', *Journal of Philosophy of Education* 14, 1: 109–16.

Flavell, J. (1978) 'Comments', in R. Siegler (ed.) *Children's Thinking: What Develops?*, New Jersey: Lawrence Erlbaum Associates.

Fluck, M. and Hewison, Y. (1979) 'The effect of televised presentation on number conservation in five-year-olds', *British Journal of Psychology* 70: 507–9.

Fodor, J.A. (1976) *The Language of Thought*, Brighton: Harvester.

Freeman, N.H., Sinha, C.G., and Condliffe, S.G. (1981) 'Collaboration and confrontation with young children in language comprehension testing', in W.P. Robinson (ed.) *Communication in Development*, London: Academic.

Freeman, N.H., Sinha, C.G., and Stedmon, J.A. (1982) 'All the cars – which cars? From word meaning to discourse analysis', in M. Beveridge (ed.) *Children Thinking Through Language*, London: Edward Arnold.

References

Frege, G. (1884) *Die Grundlagen der Arithmetik*, Breslau: Marcus; trans. J.L. Austin, *The Foundations of Arithmetic*, Oxford: Blackwell.

Garfinkel, H. and Sacks, H. (1970) 'On formal structures of practical actions', in J.C. McKinney and E.A. Tiryakian (eds) *Theoretical Sociology*, New York: Appleton-Century-Crofts.

Geach, P. (1957) *Mental Acts*, London: Routledge & Kegan Paul.

Gelman, R. (1969) 'Conservation acquisition: a problem of learning to attend to relevant attributes', *Journal of Experimental Child Psychology* 7: 167–87.

Gelman, R. (1977) 'How young children reason about small numbers', in N.J. Castellan, D.B. Pisoni and G.R. Potts (eds) *Cognitive Theory*, vol. 2, Hillsdale, NJ: Lawrence Erlbaum Associates.

Gelman, R. (1982) 'Accessing one–one correspondence: still another paper about conservation', *British Journal of Psychology* 73: 209–20.

Gelman, R. and Gallistel, C.R. (1978) *The Child's Understanding of Number*, Harvard: University Press.

Ginsburg, H. (ed.) (1983) *The Development of Mathematical Thinking*, New York and London: Academic.

Glachan, M. and Light, P. (1982) 'Peer interaction and learning', in G. Butterworth and P. Light (eds) *Social Cognition: Studies of the Development of Understanding*, Brighton: Harvester.

Gödel, K. (1931) 'On formally undecidable propositions of *Principia Mathematica* and related systems', in J. van Heijenoort (ed.) *From Frege to Gödel, a Source Book in Mathematical Logic, 1879–1931*, Cambridge, Mass.: Harvard University Press, 1967.

Grice, H.P. (1975) 'Logic and conversation', in P. Cole and J. Morgan (eds) *Syntax and Semantics 3: Speech Acts*, New York: Academic.

Grize, J.B. (1960) 'Du groupement au nombre, essai de formalisation', in *Problèmes de la Construction du Nombre (Etudes d'Epistémologie Génétique vol. XI)*, Paris: Presses Universitaires de France.

Hamlyn, D.W. (1967) 'The logical and psychological aspects of learning', in R.S. Peters (ed.) *The Concept of Education*, London: Routledge & Kegan Paul.

Hamlyn, D.W. (1971) 'Epistemology and conceptual development', in T. Mischel (ed.) *Cognitive Development and Epistemology*, London: Academic.

Hamlyn, D.W. (1978) *Experience and the Growth of Understanding*, London: Routledge & Kegan Paul.

Hamlyn, D.W. (1981) 'How does knowledge start? A reply to Pamela Moore', *Journal of Philosophy of Education* 15, 1: 137.

Hamlyn, D.W. (1983) *Perceptions, Learning and the Self*, London: Routledge & Kegan Paul.

Hilbert, D. (1904) 'On the foundations of logic and arithmetic', in J. van Heijenoort (ed.) *From Frege to Gödel, a Source Book in Mathematical Logic, 1879–1931*, Cambridge, Mass.: Harvard University Press, 1967.

Hilbert, D. (1925) 'On the infinite', in J. van Heijenoort (ed.) *From Frege to Gödel, a Source Book in Mathematical Logic, 1879–1931*, Cambridge, Mass.: Harvard University Press, 1967.

Hilbert, D. (1927) 'The foundations of mathematics', in J. van Heijenoort (ed.) *From Frege to Gödel, a Source Book in Mathematical Logic, 1879–1931*, Cambridge, Mass.: Harvard University Press, 1967.

Hughes, M. (1981) 'Can pre-school children add and subtract?' *Educational Psychology* 3: 207–19.

Hughes, M. (1983) 'What is difficult about learning arithmetic?', in M. Donaldson, R. Grieve, and C. Pratt (eds) *Early Childhood Development and Education*, Oxford: Basil Blackwell.

References

Hughes, M. (1986) *Children and Number*, Oxford: Basil Blackwell.

Hughes, M. and Grieve, R. (1980) 'On asking children bizarre questions', *First Language* 1: 149–60.

Jones, M. (1981) 'Innate powers, concepts and knowledge: a critique of D.W. Hamlyn's account of concept possession', *Journal of Philosophy of Education* 15: 139–44.

Kant, I. (1933) *Critique of Pure Reason* (ed. N. Kemp Smith), London: Macmillan.

Kant, I. (1977) *Prolegomena to Any Future Metaphysics* (trans. P. Carus, revised J.W. Ellington), Indianapolis: Hackett.

Klahr, D. and Wallace, J. (1976) *Cognitive Development: an Information-Processing View*, New Jersey: Lawrence Erlbaum Associates.

Körner, S. (1960) *The Philosophy of Mathematics*, London: Hutchinson.

Light, P. (1983) 'Social interaction and cognitive development', in S. Meadows (ed.) *Developing Thinking*, London: Methuen.

Light, P., Buckingham, N., and Robbins, A. (1979) 'The conservation task as an interactional setting', *British Journal of Educational Psychology* 49: 304–10.

McGarrigle, J. and Donaldson, M. (1974) 'Conservation accidents', *Cognition* 3: 341–50.

McGarrigle, J., Grieve, R., and Hughes, M. (1978) 'Interpreting inclusion: a contribution to the study of the child's cognitive and linguistic development', *Journal of Experimental Child Psychology* 25: 1528–50.

MacLure, M. and French, P. (1981) 'A comparison of talk at home and at school', in G. Wells (ed.) *Learning Through Interaction*, Cambridge: University Press.

Markman, E.M. (1979) 'Classes and collections: conceptual organisation and numerical abilities', *Cognitive Psychology* 11: 395–411.

Markman, E.M. and Seibert, J. (1976) 'Classes and collections: internal organisation and resulting holistic properties', *Cognitive Psychology* 8: 576–7.

Miller, P.H. and West, R.F. (1976) 'Perceptual supports for one–one correspondence in the conservation of number', *Journal of Experimental Child Psychology* 21: 417–24.

Miller, S.A. (1976) 'A nonverbal assessment of conservation of number', *Child Development* 47: 722–8.

Miller, S.A. (1977) 'A disconfirmation of the quantitative identity-quantitative equivalence sequence', *Journal of Experimental Child Psychology* 29: 180–9.

Miller, S.A. (1982) 'On the generalizability of conservation', *British Journal of Psychology* 73: 221–30.

Moore, P. (1980) 'Experience and the growth of understanding: how does knowledge start?', *Journal of Philosophy of Education* 14, 2: 261–3.

Mugny, G. and Doise, W. (1978) 'Sociocognitive conflict and structure of individual and collective performances', *European Journal of Social Psychology* 8: 181–92.

Mugny, G., Doise, W., and Perret-Clermont, A-N. (1975–6) 'Conflit de centrations et progrès cognitif', *Bulletin de Psychologie* 29: 199–204.

Mugny, G., Giroud, J., and Doise, W. (1978–9) 'Conflit de centrations et progrès cognitif, II: Nouvelles illustrations expérimentales', *Bulletin de Psychologie* 32: 979–85.

Newell, A. and Simon, H. (1972) *Human Problem Solving*, New Jersey: Prentice Hall.

Nielson, I. and Dockrell, J. (1982) 'Cognitive tasks as interactional settings', in G. Butterworth and P. Light (eds) *Social Cognition: Studies of the Development of Understanding*, Brighton: Harvester.

Opie, I. and Opie, P. (1955) *The Oxford Nursery Rhyme Book*, Oxford: Clarendon Press.

Opie, I. and Opie, P. (1959) *The Lore and Language of School Children*, Oxford: Clarendon Press.

References

Opie, I. and Opie, P. (1969) *Children's Games in Street and Playground*, Oxford: Clarendon Press.

Pascual-Leone, J. (1970) 'A mathematical model for the transition rule in Piaget's developmental stages', *Acta Psychologica* 32: 310–45.

Pascual-Leone, J. (1972) 'A Theory of Constructive Operators, a Neo-Piagetian Model of Conservation, and the Problem of Horizontal Décalages', unpublished manuscript, York University, USA.

Peano, G. (1889) 'The principles of arithmetic, presented by a new method', in J. van Heijenoort (ed.) *From Frege to Gödel, a Source Book in Mathematical Logic, 1879–1931*, Cambridge, Mass.: Harvard University Press, 1967.

Perret-Clermont, A-N. (1980) *Social Interaction and Cognitive Development in Children*, London: Academic.

Piaget, J. (1928) *Judgement and Reasoning in the Child*, New York: Harcourt Brace.

Piaget, J. (1932) *The Moral Judgement of the Child*, London: Routledge & Kegan Paul.

Piaget, J. (1950) *The Psychology of Intelligence*, London: Routledge & Kegan Paul.

Piaget, J. (1952) *The Child's Conception of Number*, London: Routledge & Kegan Paul.

Piaget, J. (1954) *The Construction of Reality in the Child*, London: Routledge & Kegan Paul.

Piaget, J. (1966) *Mathematical Epistemology and Psychology* (with E.W. Beth), Holland: Reidel.

Piaget, J. (1968a) *Genetic Epistemology*, Columbia: University Press.

Piaget, J. (1968b) *Six Psychological Studies*, Brighton: Harvester.

Piaget, J. (1969) *The Mechanisms of Perception*, London: Routledge & Kegan Paul.

Piaget, J. (1971a) *Insights and Illusions of Philosophy*, London: Routledge & Kegan Paul.

Piaget, J. (1971b) *Biology and Knowledge*, Edinburgh: University Press.

Piaget, J. (1971c) *Structuralism*, London: Routledge & Kegan Paul.

Piaget, J. (1972a) *The Principles of Genetic Epistemology*, London: Routledge & Kegan Paul.

Piaget, J. (1972b) *Psychology and Epistemology*, Harmondsworth: Penguin.

Piaget, J. (1976a) 'Postscript', *Archives de Psychologie* 44: 223–8.

Piaget, J. (1976b) 'L'individualité en histoire: l'individu et la formation de la raison', in G. Busino (ed.) *Les Sciences Sociales avec et après Jean Piaget*, Geneva: Droz.

Piaget, J. (1977) *The Grasp of Consciousness*, London: Routledge & Kegan Paul.

Piaget, J. (1978a) *The Development of Thought: Equilibration of Cognitive Structures*, Oxford: Basil Blackwell.

Piaget, J. (1978b) *Success and Understanding*, London: Routledge & Kegan Paul.

Piaget, J. (1980) *Experiments in Contradiction*, Chicago: University Press.

Pimm, D. (1981) 'Mathematics? I speak it fluently', in A. Floyd (ed.) *Developing Mathematical Thinking*, London: Addison-Wesley.

Pimm, D. (1987) *Speaking Mathematically*, London: Routledge & Kegan Paul.

Pitcher, G. (1964) *The Philosophy of Wittgenstein*, New Jersey: Prentice Hall.

Price, H. (1953) *Thinking and Experience*, London: Hutchinson.

Resnick, L. (1982) 'Syntax and semantics in learning to subtract', in T. Carpenter, J. Moser, and T. Romberg (eds) *Addition and Subtraction: A Developmental Perspective*, New Jersey: Lawrence Erlbaum Associates.

Resnick, L. and Ford, W. (1981) *The Psychology of Mathematics for Instruction*, New Jersey: Lawrence Erlbaum Associates.

Riley, M., Greeno, J., and Heller, J. (1983) 'Development of children's problem-solving ability in arithmetic', in H. Ginsburg (ed.) *The Development of Mathematical Thinking*, London and New York: Academic.

References

Rose, S. and Blank, M. (1974) 'The potency of context in children's cognition: an illustration through conservation', *Child Development* 45: 499–502.

Rotman, B. (1977) *Piaget: Psychologist of the Real*, Brighton: Harvester.

Russell, B. (1902) 'Letter to Frege', in J. van Heijenoort (ed.) *From Frege to Gödel, a Source Book in Mathematical Logic, 1879–1931*, Cambridge, Mass.: Harvard University Press, 1967.

Russell, B. (1919) *Introduction to Mathematical Philosophy*, London: George Allen & Unwin.

Russell, B. and Whitehead, A.N. (1910–13) *Principia Mathematica*, vol. I (1910), vol. II (1912), vol. III (1913), Cambridge: Cambridge University Press.

Russell, J. (1982) 'Propositional attitudes', in M. Beveridge (ed.) *Children Thinking Through Language*, London: Edward Arnold.

Schank, R. and Abelson, R. (1977) *Scripts, Plans, Goals and Understanding*, New Jersey: Lawrence Erlbaum Associates.

Scribner, S. (1977) 'Modes of thinking and ways of speaking: culture and logic reconsidered', in P.N. Johnson-Laird and P.C. Wason (eds) *Thinking: Readings in Cognitive Science*, Cambridge: University Press.

Siegler, R. (1978) 'The origins of scientific reasoning', in R. Siegler (ed.) *Children's Thinking: What Develops?*, Hillsdale, NJ: Lawrence Erlbaum Associates.

Siegler, R. (1986) *Children's Thinking*, New Jersey: Lawrence Erlbaum Associates.

Silverman, I.W. (1979) 'Context and number conservation', *Child Study Journal* 9, 3: 205–12.

Simon, H. (1962) 'An information-processing theory of intellectual development', *Monographs of the Society for Research in Child Development* 27 (2, serial no. 82).

Sinha, C. and Walkerdine, V. (1978) 'Conservation: a problem in language, culture and thought', in N. Waterson and C. Snow (eds) *The Development of Communication*, London: Wiley.

Skinner, B.F. (1957) *Verbal Behaviour*, New York: Appleton-Century-Crofts.

Smith, L. (1982a) 'Piaget and the solitary knower', *Philosophy of the Social Sciences* 12: 173–82.

Smith, L. (1982b) 'Class inclusion and conclusions about Piaget's theory', *British Journal of Psychology* 73: 267–76.

Smith, L. (1984a) 'Genetic epistemology and the child's understanding of logic', *Philosophy of the Social Sciences* 14: 367–76.

Smith, L. (1984b) 'Philosophy, psychology and Piaget: a reply to Loptson and Kelly', *Philosophy of the Social Sciences* 14: 385–91.

Smith, L. (1986) 'Children's knowledge: a meta-analysis of Piaget's theory', *Human Development* 29: 195–208.

Starkey, P. and Gelman, R. (1982) 'The development of addition and subtraction abilities prior to formal schooling in arithmetic', in T. Carpenter, J. Moser, and T. Romberg (eds) *Addition and Subtraction: a Developmental Perspective*, New Jersey: Lawrence Erlbaum Associates.

Steffe, L., von Glaserfield, E., Richards, J., and Cobb, P. (1983) *Children's Counting Types: Philosophy, Theory and Application*, New York: Praeger.

Steffe, L., Thompson, P., and Richards, J. (1982) 'Children's counting in arithmetical problem solving', in T. Carpenter, J. Moser, and T. Romberg (eds) *Addition and Subtraction: a Developmental Perspective*, New Jersey: Lawrence Erlbaum Associates.

Tizard, B. and Hughes, M. (1984) *Young Children Learning*, London: Fontana.

Tough, J. (1976) *Listening to Children Talking*, London: Ward Lock Educational.

Trabasso, T., Isen, A., Dolecki, P., McClanahan, A., Riley, C., and Tucker, T. (1978) 'How do children solve class-inclusion problems?', in R. Siegler (ed.)

References

Children's Thinking: What Develops?, New Jersey: Lawrence Erlbaum Associates.

Walkerdine, V. (1982a) '2p doesn't buy much these days: learning about money at home and in early mathematics education', in V. Walkerdine *Linguistic and Cognitive Development in Home and School Contexts*, London: SSRC Final Report.

Walkerdine, V. (1982b) 'Relational terms in everyday social practices: *more* or *less* reconsidered', in V. Walkerdine *Linguistic and Cognitive Development in Home and School Contexts*, London: SSRC Final Report.

Walkerdine, V. (1982c) 'The interaction between size and family terms in home, school and experimental contexts: Part One', in V. Walkerdine *Linguistic and Cognitive Development in Home and School Contexts*, London: SSRC Final Report.

Walkerdine, V. (1982d) 'The interaction between size and family terms in home, school and experimental contexts: Part Two', in V. Walkerdine *Linguistic and Cognitive Development in Home and School Contexts*, London: SSRC Final Report.

Walkerdine, V. (1982e) 'Pre-school, pre-mathematics: some thoughts towards a critique', in V. Walkerdine *Linguistic and Cognitive Development in Home and School Contexts*, London: SSRC Final Report.

Walkerdine, V. (1982f) 'Mathematics at home and at school: an overview', in V. Walkerdine *Linguistic and Cognitive Development in Home and School Contexts*, London: SSRC Final Report.

Walkerdine, V. (1982g) 'From context to text: a psychosemiotic approach to abstract thought', in M. Beveridge (ed.) *Children Thinking Through Language*, London: Edward Arnold.

Walkerdine, V. and Sinha, C. (1978) 'The internal triangle: language, reasoning and the social context', in I. Markova (ed.) *The Social Context of Language*, London: Wiley.

Wilkinson, A. (1976) 'Counting strategies and semantic analysis as applied to class inclusion', *Cognitive Psychology* 8: 64–85.

Wittgenstein, L. (1953) *Philosophical Investigations*, Oxford: Basil Blackwell.

Zoetebier, J. and Ginther, T. (1978) *Social Interactie en Cognitive Ontwikkeling*, Tilburg: Katholieke Hogeschool, Subfaculteit Psychologie.

Index

Abelson, R. 152
abstraction: 45, 54–5, 64–5, 70–3, 82, 124; *see also* reflective abstraction
abstractionism 82
accommodation 48–50
action schemes: coordination of 51–3; internalization of 44, 55–9; and objects 50–5
Aristotle 13
arithmetic: context (*q.v.*) of doing 147–50; conventions of 126–9, 153, 158; foundations of 12–18; social practices (*q.v.*) of doing 126–9, 158–9, 161–2, 174–8, 185–6; *see also* mathematical thinking; mathematics; number
assimilation 48–50

background to Piaget's theory 9–27; mathematical propositions 10–18; mathematics and genetic epistemology 18–27
Balfour, G. 133–4
Barker, S. 13–14, 17, 34
Barnes, D. 168
Beth, E.W. 9
Blank, M. 88, 137, 170
Boole, G. 19
Brainerd, C. 82, 188–9n
Bruner, J. 93
Bryant, P. 5–6, 89–98, 108, 131–2, 190n; invariance principle and conflicting relative code cues 92–7; perceptual cues and relative number code 90–2; theory 90, 97–8

Cantor, G. 12–15, 22
Carpenter, T. 189n
Case, R. 5, 119–30, 190–1n; instruction theory 124–9; 'neo-Piagetian' analysis 119–30; task analysis 121–4
Chomsky, N. 43, 46, 67, 76
class inclusion: 60–1; tests 88–9, 134–6, 139–41, 144–5; Piaget's theory 30, 36–42, 81
classification 29–30, 37–8, 80–1

classroom: arithmetic 174–8, 184–6; social practices of 170–4
Cobb, P. 7, 131, 150–2
codes for number, relative 90–7
concepts formation of 55–9, 68–9; number (*q.v.*) 28–42; self and objects 50–5
concrete-operational thought, structure of 36–9, 56–7, 60–1, 80–1
Condliffe, S.G. 181
conflict: Bryant's theory 93–7; Case's theory 122–4; inter-individual *v.s.* intra-individual 111, 115; perception of in Piaget's theory 49–50, 53–4, 59, 61, 66, 111, 124; sociocognitive 110–19
conservation 29–30, 61, 80–1, 92–3; and one–one correspondence 103–7; task analysis of 121–4; tests 88, 92–7, 102, 117–19, 133, 136–7, 143–4, 180
context of doing arithmetic 147–50; conventions of 126–9; social practices of 159, 161–2, 185–6; understanding the 150–9
Cooper, D.E. 189–90n
Corran, G. 163, 185–6
counting: Gelman and Gallistel 99–102; Steffe *et al.*'s typology of 151–6
Crowther, R. 161, 181–4
cues, and relative codes 90–7

Dearden, R.F. 148–50
De Corte, E. 191n
Dockrell, J. 137
Doise, W. 5, 110–19, 130
Donaldson, M. 6, 88, 132–8, 141–5
Durkin, K. 161, 165, 181–4

Edwards, D. 173
Elliott, R.K. 190n
empiricism 21–2; Piaget's criticism of 44–6, 62–3
epistemology 9, 19–21, 26, 41–2, 47, *see also* genetic epistemology
equilibration 47, 48–50, 69; action

200